THE SOURCE

THE SOURCE

HOW RIVERS
MADE AMERICA
AND AMERICA
REMADE ITS RIVERS

Martin Doyle

W. W. NORTON & COMPANY
Independent Publishers Since 1923
New York | London

For information about permission to reproduce selections from this book, write to
Permissions, W. W. Norton & Company, Inc., 500 Fifth Avenue, New York, NY 10110

For information about special discounts for bulk purchases, please contact
W. W. Norton Special Sales at specialsales@wwnorton.com or 800-233-4830

Manufacturing by LSC Communications Harrisonburg
Book design by Chris Welch
Production manager: Lauren Abbate

ISBN: 978-0-393-24235-5

W. W. Norton & Company, Inc., 500 Fifth Avenue, New York, N.Y. 10110
www.wwnorton.com

W. W. Norton & Company Ltd., 15 Carlisle Street, London W1D 3BS

1 2 3 4 5 6 7 8 9 0

FOR JORDAN, STUART, AND EVELYN—MY BOW PADDLERS.

CONTENTS

PART FOUR: REGULATION

PART FIVE: CONSERVATION

INTRODUCTION

America has more than 250,000 rivers—over 3 million miles—dissecting the fields, cities, and forests of the nation. These aren't just any rivers; these are world-class rivers. The Mississippi watershed alone drains the rain and snow that falls on over a million square miles, generating 390 billion gallons of water per day, enough over the course of a year to cover the entire United States with several inches of water. The falling of these rivers over the continent's topography generates tremendous power: the Grand Coulee Dam on the Columbia River is the largest power plant in the United States, yet it makes up only a third of the power production of all the dams just on the Columbia. And America's rivers span enormous climatological gradients. The Colorado most notably brings runoff from the snow-rich Rockies through thousands of miles of desert, providing the water for nearly 30 million people and more than 1.8 million acres of irrigated land and allowing the production of 15 percent of the nation's crops in one of the driest regions on Earth. Anyone in the United States who eats a salad in the winter is enjoying lettuce grown with Colorado River water. Rivers are the defining feature of America's landscape.

Rivers have shaped the basic facts of America. Most simply, of

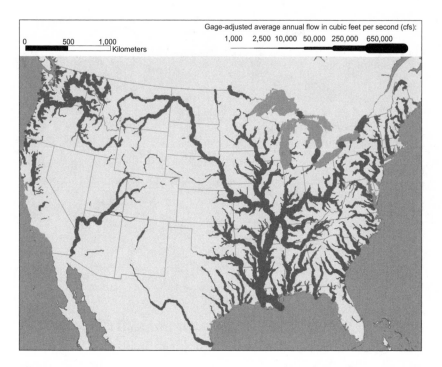

Gage-adjusted average annual flow in cubic feet per second (cfs):

The rivers of the United States scaled by flow.

course, our borders themselves are typically set by rivers—the Mississippi, the Ohio, the Red, the Columbia, and the Colorado all form borders between states while the Rio Grande separates the United States from Mexico. Sixteen states were named after rivers, along with over 150 counties.

Rivers have also shaped where we live. When European colonists arrived on the Eastern Seaboard of America, they established their farms in the coastal plain, where land was flat and the sluggish rivers, streams, and sloughs were easily plied. As newly arriving settlers moved farther inland, they confronted a geologic peculiarity where the Atlantic coastal plain runs up against the Piedmont—the Fall Line. Upstream of the Fall Line, rivers are entrenched in narrow, steep-walled valleys interspersed with rapids. Downstream are the easily traveled rivers of the coastal plain. At the Fall Line lie rapids or even waterfalls. For anyone trying to move a bale of cotton or a bushel of wheat from the interior by boat, the Fall Line would mark the end of the line where canoes moving downstream would have to be unloaded, portaged, and reloaded. To move goods upstream from the Atlantic, cargo would have to be shifted from larger coastal or ocean vessels to land transportation or smaller canoes or rafts to transport upstream.

All this exchange of goods meant that both goods and money were changing hands: foreign merchants with ships interspersed with country bumpkins and their wagons. In Fall Line ports, entrepreneurs and job seekers alike smelled opportunity. And at the Fall Line ports of the eighteenth century, villages and towns began forming and growing as nascent hubs of commerce, the centers of the emerging market economy. These random places at the intersection of rivers and the Fall Line became natural places for immigrants to settle and cities to grow: Richmond on the James, Washington on the Potomac, Trenton on the Delaware, and nearly all of the other cities peppering the Eastern Seaboard. That the major cities of the eastern United States, and particularly the state capitals, are located at the Fall Line is a simple indicator of how subtle the effect of the physical landscape can be in shaping the demographic landscape.

Cities of the Midwest, the Deep South, and the West are little different. Many cities of America's interior lie at important river confluences: Pittsburgh where the Monongahela and Allegheny join to form the Ohio, Kansas City at the Missouri and Kansas, and Sacramento at the Sacramento and American rivers. Cities of the Deep South, where rivers are plentiful, are typically situated to avoid water—or at least floods. Memphis, Vicksburg, and Natchez all sit up on a bluff overlooking the Mississippi River from a safe elevation yet close enough to be a bustling port for nineteenth-century steamboats or twenty-first-century barge tows. Two of the oldest Mississippi River cities, Saint Louis and New Orleans, likewise sit on slight topographic rises—natural levees—that have kept their residents above the hydrologic fray for centuries.

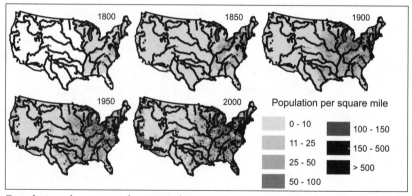

Population density in the United States has traditionally been bounded by and clustered around rivers.

Other centers of America's riverine society are often less populated but no less formative: they are hubs of energy. Where rivers are constricted to narrow valleys and canyons, or where they cascade over especially powerful waterfalls, their physical power has been harnessed against the rapidly changing backdrop of evolving technology. Settlers of the eighteenth and nineteenth century built their villages around small dams powering waterwheels; twentieth-century engineers and federal agencies built monumental concrete dams and countless turbines to generate power that would be sent through an ever-sprawling electric grid to distant cities and far-flung industries. The power of the Susquehanna River was as essential to grinding colonial grain as the Merrimack River was to spinning the fabric of New England textile mills; the Columbia River was as crucial in extracting aluminum for World War II bombers as the Tennessee River was in refining the uranium those bombers carried. The evolution of technology has allowed rivers to power the rise of America from colonial backwater to industrial juggernaut.

Demographics, technology, and the economy—all these aspects of America's history have played out on a landscape defined by rivers. The simple geologic process of water eroding bits and pieces of sediment has resulted in physical features that have shaped, and continue to shape, the events of modern society.

But rivers have also shaped the very ideas of what America should be. Is a standing, permanent army needed, or might a loose collection of state militia be sufficient? The need for fortifications at the harbors of interstate river ports made a standing army both acceptable and necessary. Which level of government should regulate commerce? River-borne commerce on interstate rivers drove the new republic to have a single national economy rather than a collection of independent state economies. How big should the nation be? The Louisiana Purchase—the acquisition of the entire Missouri River basin—changed the basic geography of America into a vast westward empire. In so doing, it gave enough space for the quintessential frontiersmen and pioneers, along with the idea of manifest destiny. How active should the national government be? Floods of biblical proportion in the early twentieth century, when combined with a nationwide economic depression, forced the nation to accept a far larger role for the federal government than had been tolerated before. Many of the issues in American society have arisen from or been fought over the fact of our riverine republic.

Meanwhile, as its rivers have formed American society, Americans have changed their rivers. We have drained, straightened, leveed, and dammed them; polluted and befouled them; cleaned up and restored them.

Tracing the history of America's rivers thus requires disentangling the political, demographic, technological, and economic contexts that predominated in specific eras while simultaneously understanding the hydrological events constantly at work to naturally re-carve rivers. Amid these events and contexts are the people who embody the moments, the decisions, and the ideas as they evolved. Some are highly visible—like David Lilienthal, father of the Tennessee Valley Authority, and Andrew Humphreys, visionary of Mississippi River flood control. Others are less known but equally transformative—

the engineers, geologists, towboat captains, and textile mill workers whose lives and labor have been mixed with the flow of rivers.

Reading America's rivers is akin to reading a palimpsest—one of those ancient manuscripts from which text was washed off or scraped away to make space for new writings, under which each previous layer was visible but not fully distinct from those that preceded it. America's rivers are likewise a series of decisions on top of decisions, events on top of events, ideas on top of ideas. Each has its own context, but none can be completely separated or even distinguished from the layers above or below it.

This book works through these contexts and layers in five parts. Part One looks at rivers through the lens of federalism, the most important concept affecting how American governance was to be structured, and how this peculiar structure shaped river navigation and flood control. The division of water between people and groups of people in this country has been influenced by our thinking about sovereignty and property rights, the focus of Part Two. From there, Part Three focuses on our ever-evolving view of taxation—how the government will get revenue and then spend it. Along with the ability to tax comes the ability to regulate the economy, and Part Four traces the history of the changing notions of regulating a specific part of the economy—power. Finally, Part Five looks toward a hopeful future of environmental conservation by considering the nation's evolving ideas of environmentalism.

How any society interacts with its particular environment is a product of these types of underlying ideas that shape the people in that society. These ideas are as formative in how we transform the landscape as they are in how we structure the government or regulate the economy. This book is about how our ideas have shaped the rivers of America—and how the rivers have shaped our ideas.

PART ONE

FEDERALISM

CHAPTER 1

Navigating the Republic

On September 24, 1784, George Washington was in the north-west extremes of Virginia along the west flank of the Allegheny Mountains, over two hundred miles from his home in Mount Vernon. The region was at the headwaters of the Monongahela River, which combines with the Allegheny to form the Ohio River. Throughout the previous weeks, Washington had been crisscrossing the mountainous region to check on his landholdings in the "West"—land that was to the west of the Appalachians, recently ceded to the United States as part of the Treaty of Paris. Essentially, Washington was a land speculator. Like many others in America, he had built his fortune by amassing land; and much of the land he and these other prospectors had acquired was in the western territory of the Ohio River valley. The land was enormously fertile; but without the ability to get to the land, and to get the products out easily, the land was largely worthless. Its value hinged on whether the barriers of first the Fall Line and then the more formidable Appalachians could be conquered. So Washington spent the years between his Revolutionary War victory and his presidency trying to find a way to connect the Atlantic to the Ohio River.

Washington's motives for seeking a path through the region were not solely self-interested. He had grown increasingly concerned about the myriad complexities that the Appalachian Mountains introduced to the new republic, as had Thomas Jefferson. In 1784, Jefferson foresaw that the link to the "western country" was a key to commercial expansion of the new republic. But Jefferson was also keenly aware of the regional differences between the North and South, already present in the earliest days of the Republic. He foresaw the potential for division in the competition that would inevitably occur over which state or region would control the route through the barrier of the Appalachians, which region could capture "as large a portion as we can of this modern source of wealth and power." Just as European powers had competed to cross the barrier of the Atlantic quickest and most efficiently to establish the hub of commerce, a geographic competition between those based along the Potomac, the Hudson, and the Mississippi was under way for access to the Ohio River. Jefferson was certain that "there would be a rivalship between the Hudson [in the North] and the Potomac [in the South] for the residue of the commerce of all the country westward of Lake Erie."[1] The race for tapping agricultural wealth to the west of the mountains was thus potentially a race between the North and the South.

Furthermore, the economic vitality of the infant United States was at stake. This economic competition was not confined to different regions of the United States; it also involved foreign powers. Commodities flowing down the Ohio to the Mississippi would pass through New Orleans, which was then under Spanish control. Those potentially being moved to the North would pass through the Great Lakes and on to the St. Lawrence River and Canada, then under British control. If the barrier of the Appalachians remained, the American Republic would be economically isolated from the burgeoning west, and European powers would claim the most fertile regions of what was then known of the continent. America would be a vast coastline with few inland opportunities, ensconced within an impenetrable mountain chain.

The peculiar geography of U.S. rivers had yet another, more subtle effect. The predominant east–west axis of rivers of the Atlantic Seaboard meant that no significant rivers passed through both Northern free states and Southern slaveholding states. Transport of goods by rivers, the predominant mode of commercial traffic at the time, was conducted within single political ideologies; in many ways, river commerce served to further divide the Republic instead of uniting it. Jefferson saw that, were a portage or canal constructed that linked the western region to New York, the North would gain commercial supremacy and the primary flow of commodities from the West would pass through only free states and on to free Britain. The unique political economy of the slaveholding South would be isolated. Binding the separate ideologies together required binding them together economically, which hinged on some north–south river navigation in addition to cross-mountain access to the West.

All of these concerns were on Washington's mind when he made his trek out West in 1784. He traveled to the western territory in the most efficient way possible at the time: up the Potomac River as far as Cumberland, Maryland, and then by foot overland to the Ohio regions. The Potomac River crosses the Fall Line at a place called Great Falls, a description well deserved because that section of river is more like a waterfall than a rapids. Upstream of Great Falls, the Potomac is passable by boat but interspersed with rapids and difficult to navigate. Upstream of Cumberland, Washington would have had to go without boats, up and over the Appalachians, and then back into the streams and rivers of the upper Monongahela, which were no more easily passed than the upper reaches of the Potomac. It was a hard trip both on water and on land: by water meant frequently getting out and carrying the boats around the rapids; by land meant carrying all the supplies and fording the incessant rivers, streams, and creeks. Washington went by land, traveling 680 miles from September 1 to October 4, 1784.

The focus of Washington's trip through this region was surveying, and his diary contains few impressions of the people and landscape but many cold, hard facts of geography. His record of the trip is filled with details of fording rivers and noting distances for portages:

> That [portage] from the fork of Monongahela & Cheat, to the Court House at Morgan Town, is, by Water, about 11 miles, & from thence to the West fork of the former is 18 More. From thence to the carrying place between it and a branch of the little Kanhawa, at a place called Bullstown, is about 40 Miles, by Land—more by Water and the Navigation is good.[2]

At the time of this surveying trip, the 52-year-old Washington was both a rich plantation owner and a war hero. He was tall and regal—and even better, aloof. It was difficult for him to tromp around the backwoods on the edge of the Republic without drawing attention. And when he asked for opinions, he inevitably drew a crowd, as he did on a particular September night at George's Creek on the west flank of the Appalachians. The crowd gawking at Washington was crammed into a tiny cabin—the local land agent's office and home, it was about the size of a small college dorm room. Beyond being in awe of the general, the backwoods hunters and surveyors had gathered at Washington's request. These were the men who knew the landscape and how to get around in it. They were at the cabin to give their opinions on the best possible routes for a portage road connecting the Potomac to the Ohio.

The group deferred to Washington, whose cold, distant demeanor and legendary reputation were well known; each man spoke only when acknowledged. The oddball of the group was a young Swiss immigrant who spoke English with a strong French accent. Like Washington himself, he was in the area to survey, searching for land that was situated where travelers must pass when—inevitably—a road was built connecting the Potomac to the Ohio. Buying such land ahead of time and in just the right place would enable him to make enormous profits. Brilliant and impetuous, the young man finally grew frustrated with both the group's and Washington's equivocation over possible routes. He interrupted the general to blurt out, "Oh, it is plain enough," as he pointed to the route that would be best.

An awkward silence followed as the men in the cabin stared at the young surveyor. Washington put down his pen and, offended at being interrupted, contributed his own cold stare. But later, when discussion resumed, the general suddenly stopped and put down his pen. He turned to the Swiss immigrant and commented, simply, "You are right, sir."[3]

And so the young Albert Gallatin must have felt quite satisfied as he wrapped himself in buffalo skins that evening, dozing in a crowded cabin where General George Washington also slept. The next morning Gallatin turned down an offer from Washington to oversee the General's western landholdings, and Washington headed east over the mountains toward Mount Vernon. It would be another five years before Washington was president. It would be 17 years before Gallatin became Secretary of Treasury and 21 years before Lewis and Clark stood at the Missouri River to christen one of its three tributaries as the Gallatin River.

When Washington returned from his western explorations, he was eager to take the first step to link the East and the West: establishing viable navigation on the Potomac River. Doing the necessary work on the river required the formation of a company that could work on an interstate river. But such a company was difficult to establish due to the existing political climate and structures. In 1784, the United States was governed by the Articles of Confederation. Each of the states functioned independently—and the Potomac River forms part of the border between two states. Getting congressional approval on an agreement between states for work on an interstate river like the Potomac would be nearly as arduous as establishing a treaty between separate countries to do work on an international river like the Rio Grande or the Columbia.

Washington, anxious to get his company off the ground, was exasperated with the unwillingness of Congress—or its inability under the Articles of Confederation—to recognize his already-negotiated interstate deal; the lawmakers' intransigence was his continued economic loss. He vented to Jefferson that negotiating a treaty for interstate commerce was impossible given the "incertitude which prevails in Congress, and the non-attendance of its members."[4]

The following year, in 1785, delegates from Maryland and Virginia—those states bordering the Potomac—gathered in Washington's home at Mount Vernon. They met to work out an arrangement for commerce on the river that would become known as the Mount Vernon Compact. They came up with plans for governing tolls, fishing, and other commercial interests on the Potomac, as well as navigation on the Pocomoke River and Chesapeake Bay. All of these were steps in the right direction, but in arriving at the agreement that would address the problems Washington's company confronted, the representatives noted that a whole range of additional issues remained unaddressed. After approving the agreement, which was in turn approved by their respective legislatures, one of the participants—James Madison, from Virginia—suggested holding another meeting on interstate commerce issues and inviting representatives from all the states.

In September 1786, this follow-up conference was held in Annapolis, Maryland. Once again, Washington was unable to push forward his Potomac Company. In fact, reconciling many of the conflicts that arose out of interstate commerce proved difficult, primarily because of the limitations imposed by the existing Articles of Confederation. Annapolis raised many additional issues but did not resolve them. The delegates were frustrated, but saw advantages to broadening the conversation even further, so they issued a call for yet another meeting. This third meeting would be in 1787 in Philadelphia, and it would be a full constitutional meeting. With interstate commerce as its initial purpose, the Philadelphia convention tackled a wide range of issues, eventually even setting aside the Articles of Confederation to develop an entirely new set of governing principles, which yielded the U.S. Constitution. As Teddy Roosevelt would argue over a century later, the Constitution had its roots in interstate river navigation.

The Erie Canal Village Museum in upstate New York is just east of the Rome Sports Hall of Fame Museum. On this hot July day, neither place is crowded. The parking lot at the canal museum is more grass than gravel. The sign is off-kilter and looks as if it was last painted 20 years ago. The weeds at the edge of the property are greedily encroaching on the museum grounds, laying a quiet biotic siege.

The three people working at—or at least loitering in—the museum are startled to see a visitor, but enthusiastically answer my questions about local history. They are a wealth of information about the canal, the boats that traveled on it, and the small villages and cities that developed alongside it. An anecdotal history of the canal in the mid-nineteenth century is forthcoming, but a seemingly simple question flummoxes them: How much commercial traffic moves on the canal system now?

"Wow, that's a good question."

"Now, I don't know that."

"Can't say if I've ever seen a boat with a real barge."

"What do you mean by *commercial*?"

"If you find out anything, can you call us and tell us?"

"Go ask one of the lock tenders on the Barge Canal. They'll know everything."

The modern-day Erie Canal is a conundrum, and for a feature of the early American landscape that so profoundly affected the course of history, it is surprisingly hard to find. To understand the Erie Canal demands an appreciation of topography. The canal route really starts at Troy, New York, just upstream of Albany, where the Mohawk River flows into the upstream end of the easily navigated portion of the Hudson River. From there, the route heads directly west along the valley of the Mohawk River, which is remarkable for its flatness—the river rises surprisingly little upstream of Troy as it passes through miles of rolling hills in upstate New York. As you head toward the source of many rivers, their valleys become deeper and narrower, forming a distinct V in the landscape. But as you continue west toward the source of the Mohawk, its valley grows less obvious. West of Rome, the Mohawk valley fades into a broad, flat landscape.

The canal itself sometimes seems to disappear altogether, or at least become almost impossible to find. This is partly because the Erie Canal is now really two canals: the original canal, which was replaced in 1918 by the modern canal, known as the Barge Canal. With a bit of looking, a lot of detailed map reading, and a few missed turns, it is possible to find—really, stumble onto—the original Erie Canal. In many places today, the original canal is little more than an algae-covered gully—no bigger or more attractive than any drainage ditch on one of the sprawling farms nationwide.

It is far easier to track down the Barge Canal. Still, it's a far cry from the heyday of the canal, when you could have located it most easily by following crowds of people into a bustling downtown hub of villages, towns, or cities. Nowadays you have to use GPS or the small signs along backroads to find the canal.

Like all the buildings along the modern Barge Canal system, the small house and power building at Lock 20 are painted dark blue with bright yellow numbers. These are the same colors as the New York State Thruway, for good reason: the Thruway authority owns, operates, and pays for the canal system. Strangely, and adding to how confusing all things are about the modern Erie Canal, the Thruway charges tolls for cars and trucks, but the canal does not charge the barges and boats.

Today's lock tender, or lockmaster, is busy reading the newspaper. At my request he kindly takes the time to describe what typically passes through the canal: "Yachts; motorboats; fishing boats. Normal stuff." When I ask about commercial traffic, his eyes light up. While

The route of the Erie Canal in New York State.

he's never locked through a real barge, he mentions "a few hundred tons of oats" expected to come through from Canada sometime this summer.

At another of the three dozen locks along the Barge Canal, about ten miles away, the two lock tenders are generally hostile to my questions. They are in the middle of locking through four house-boat/yachts—a lot of excitement for the canal these days. At two other locks along the canal, there is no traffic to manage, but there is a great sense of anticipation about the barges of "tons and tons of oats" that will pass through sometime in the future. And at another lock farther west, the lock tender, who had previously worked for years in the toll booths along the Thruway, has also heard about the legendary oats coming from Canada—"a hundred barge loads or so"—but he's not sure when they will reach his particular lock.

Later, over the phone, an official representative of canal operations for the New York Thruway confirms that she has also heard about the loads of oats coming from Canada: "Several thousand tons, maybe more." The call is hard to hear over the constant whine of trucks from the nearby Interstate-90, all paying tolls to the State of New York, which manages the Thruway and subsidizes the yacht-facilitating canal.

The modern, derelict condition of the Erie Canal belies its formative role in the nation's early years as its transportation ("navigation") system. As we've seen, the problem of navigating rivers for interstate commerce motivated the states to convene the Constitutional Convention. The Constitution that resulted was, and is, not easily interpreted. Rather than being conceived of as a coherent set of commandments, it was shaped by a series of compromises among widely differing factions. After thirty-nine delegates signed the Constitution in 1787, they had to convince their respective states to agree to adopt the Constitution as the new binding government. But their work was not finished; they also had to clarify the meaning of the Constitution to their constituents and fellow delegates. Three then anonymous commentators took up the task: Alexander Hamilton, John Jay, and James Madison, who collectively adopted the pen name "Publius."[5] They began writing commentaries for newspapers, eventually penning eighty-five such articles in a series called *The Federalist*, now known as the *Federalist Papers*. The significance of the *Federalist Papers* in our time is that they articulate how the framers understood the Constitution when they wrote it and, just as important, how the people of the United States understood the Constitution when they ratified it.

A central idea proposed in the Constitution involved forming a stronger, national government and dissolving the existing weak union between confederated states, which were linked more by an alliance than a government. That national government—what we call the federal government—would be central for some functions while other functions would be left to the individual states. At the time, many people felt that America was simply too large geographically for a single national entity to govern any functions. It was not clear how people from southeastern Georgia could be stitched together in common interest with people from northwestern Massachusetts. John Jay, in *Federalist* No. 2, argued that the nation was in fact a union rather than a group of disparate entities and that this coalition was enabled by sharing a common language as well as a network of rivers:

> It has often given me great pleasure to observe that the independent America was not composed of detached and distant territories, but that one connected, fertile, wide-spreading country was the portion of our western sons of liberty. Providence has in a particular manner blessed it with a variety of soils and productions and watered it with innumerable streams for the delight and accommodation of its inhabitants. A succession of navigable waters forms a kind of chain round its borders, as if to bind it together; while the most noble rivers in the world, running at convenient distances, present them with highways for the easy communication of friendly aids and the mutual transportation and exchange of their various commodities.[6]

Rivers would allow movement between disparate states, and the commercial trade they made possible would bind the residents of different states.

Indeed, it was in the interest of facilitating commerce between the various states that people generally agreed on the need for a new form of government. In *Federalist* No. 11, Hamilton said, "The importance of the Union, in a commercial light, is one of those points about which there is the least room to entertain a difference of opinion, and which has, in fact, commanded the most general assent of men who have any acquaintance with the subject."[7]

The first twenty-five of the *Federalist Papers* focused on articulating the need for a stronger central government and on the types of functions, like commerce on interstate rivers, that could not be readily accomplished without such a government. From there, the *Federalist Papers* articulated what the central government could not do, in the process establishing a central feature of the Constitution—federalism. That is, while the national government would take up some functions, it would leave others to the state governments. States were to be the locus of power under the Constitution, and they were to be the government entity initiating projects and programs. By necessity, a key function that state governments immediately took on was putting their rivers to work for navigation.[8]

At the close of the eighteenth century, settlers were moving beyond the Fall Line and into the Piedmont, where farms began appearing up to the foothills of the Appalachian Mountains. Every city and town along the Fall Line wanted to keep upstream stretches of the rivers through the Piedmont navigable, not least to ensure the ability to easily move people and products from these farms to harbors at the coast, and vice versa. Thus these cities, and their states, needed a canal to pierce the barrier of their local waterfall or rapids. Canal companies and river navigation companies were the entrepreneurial answer. Like twenty-first-century dot-coms, these companies were the investment of choice to change the world and get rich while doing it.

Such companies were formed by private investors and given the right by their state government to build canals and clear rivers for navigation. As an incentive, the companies also got a monopoly over that particular reach of river and the right to collect tolls on it. River and canal companies were among the earliest forms of publicly recognized private corporations, an early test of how the infant American government would interact with the economy: state governments were the sole government entity involved in these new ventures, for the Constitution at the time was interpreted to hold that the national government did not have an interest or function in such private enterprise.

The first river navigation companies had modest ambitions: keeping their river cleared of logs, sandbars, and any other blockages. Often formed by local landowners, who had great personal motivation for the rivers to be navigable, the companies sprouted up along just about every river in a state. The North Carolina legislature, for instance, recognized the Cape Fear Navigation Company, the Neuse River Company, the Tar River Company, and the Roanoke River Company; thus, all four of the state's primary rivers were controlled by private companies formed to facilitate navigation. Canal companies were a bit more ambitious, for they were more capital and expertise intensive and most often focused their attention on getting traffic around the rapids at the Fall Line.[9]

State governments supported this development of their internal economies. They invested in these new companies as stockholders without becoming involved in the direct management or operations. In passing legislation that allowed river navigation companies to clear rivers or canal companies to construct locks, legislators were developing a restrained role for government: they would financially support commerce and trade but would not do the actual work nor directly own the public works. Before 1793, eight states had incorporated thirty canal companies; by 1823, New Hampshire alone had chartered twenty.[10]

Almost immediately, the early companies and their sponsoring states found that the costs involved in designing and building canals and locks were massive, as were the chronic costs of keeping rivers clear for navigation. What's more, costs nearly always exceeded expectations; timelines were habitually extended; and very little of what was planned actually worked. Even when projects were successfully completed, floods often damaged locks, requiring weeks or months of repairs and frustrating commercial traffic to the point of diverting it to primitive toll roads (called turnpikes) for horse-drawn wagons.[11] All of these costs and efforts centered on just getting over the Fall Line; the greater financial opportunity—and the more formidable challenge—would be in getting through the Appalachians and linking the Atlantic economy with that of the West: the Great Lakes and the Ohio River basin. But the mountains created seemingly impossible tasks for entrepreneurs and state governments. In their stubborn efforts to link the Potomac to the Ohio, the Chesapeake and Ohio Canal Company faced a route estimated to require 246 locks and a 4-mile tunnel piercing the divide at an elevation of 1,900 feet. Likewise, over 1,400 feet of elevation gain were needed for the path through Pennsylvania along the Allegheny Portage route, which would require over 174 locks. Other west–east paths, like those along the James River in Virginia, were even worse.[12] Although stone turn-pikes could be constructed for $5,000 to $10,000 per mile through mountainous regions, and most canals in the Piedmont might cost $20,000 to $30,000, the Chesapeake and Ohio Canal, which depended greatly on locks as it gained elevation, cost over $60,000 per mile as it moved into higher reaches.[13]

When private capital became scarce, state governments pushed the projects along by continuing to invest as stockholders. For states to do this, even though the exorbitant costs exceeded their immediate revenues from taxes, they had to take on debt. They sold state-backed bonds to generate their own capital, some of which they invested in canal and river companies. The revenue from the canal and river tolls—paid directly to the companies and then back to the sharehold-ing states—became a key source of revenue for state treasuries, making the financial success of states inseparable from the success of canal and river companies. State economies and river traffic became irreversibly intertwined. Meanwhile, the national government's finances remained outside the fray of river traffic. All taxation and spending related to river clearing and canal building was left, practically and politically, to the states, thus separating the national government from the state gov-ernments with respect to the frenzy of ongoing river projects.

One state—New York—was different. At its highest point crossing toward the Great Lakes, the Erie Canal route rises less than six hun-dred feet above the Hudson River at Albany and required just over eighty locks. The heavy lifting of getting up and over the spine of the eastern continent—what engineers, shovels, reservoirs, and locks had to accomplish in Pennsylvania, Maryland, and Virginia—had been done largely by the vagaries of glacial erosion in New York.

Most of the river valleys on the eastern flank of the Appalachian Mountains had been formed by very gradual river-dominated erosion. These valleys are typically steep and narrow in their headwaters, followed by occasional rapids or falls through the Piedmont until their rivers wander through shoaled coastal plains. But along the Hudson River and its tributary the Mohawk, glacial erosion scoured existing river valleys far deeper and wider than those made by rivers alone in the warmer, unglaciated south. As glaciers receded and sea level rose, the Hudson Valley became a tidal inlet, creating a deep river that was easily navigated. Its tides extended all the way to Troy—a full 160 miles inland—where the water surface at high tide is only 10 feet above sea level. So Pleistocene glaciation did a marvelous job of digging the Hudson River deep; in the Mohawk River valley the same glaciation scraped away bedrock and, as the glaciers receded, deposited massive layers of soft, easy-to-excavate sediments across the valleys. This glacially smoothed gap is an exceptional, peculiar landscape that is unusually perfect for a canal.

The Erie Canal was all that was needed to penetrate the barrier between the Atlantic and the burgeoning West. It funneled traffic and commodities along its northern pathway instead of through the South. When initially completed, the canal was only 4 feet deep and 40 feet wide. It allowed two horses on the parallel towpath to pull over thirty tons of freight at the steady clip of 4 miles per hour. Costs to move freight dropped from $125/ton before the canal to less than $6/ton after. Within a year of its completion, over seven thousand boats were operating on the canal. The canal was part and parcel with industrializing the North, and the once obscure towns along the canal—Syracuse, Utica, Rochester, and Buffalo—became hubs of nineteenth-century manufacturing and industrialization.

While other canals in other states were struggling to keep only a handful of locks operating, the Erie Canal was successful even before it was completed in 1825—a pivotal year for rivers in the United States. A year before the canal opened, its revenues exceeded its loan interest by over $400,000. Villages on the river turned into towns as even the partial opportunities offered by the incomplete canal were vastly superior to the hazardous and hit-or-miss conditions of the Potomac and other canals or portage roads to the south. The swampy village of Syracuse sat on isolated lowland in the middle of the state and had an anemic population of 250 in 1820 when the Erie construction crews approached it. Five years later, Syracuse's population had more than doubled to 600; it then leaped geometrically to over 11,000 by 1830 and to over 22,200 by 1850. Despite few other resources or geographic advantages, Syracuse managed to flourish as a hydraulic crossroads of northern economies where the Erie Canal provided continuous passage to Lake Erie and the Oswego Canal headed north to Lake Ontario.[14]

The canal also established New York City as the entry point for immigrants heading toward America's interior. An 1836 letter from Karl Brunnhuber, living in Ohio, to his brother Frederick back in Germany shows the role of the Erie Canal and tells of the hopscotching immigrants had to go through to make their way to the American interior:

> If you can leave Erlangen by mid-June at worst, you will arrive in time to help with the harvest. . . . Go first to Albany then buy a passage on a line boat on the Erie Canal. A passenger boat would be faster but a line boat is cheaper and has room for the woodworking tools I want you to bring. Try to find a clean boat that is not too crowded. At Buffalo you must get a steamboat across Lake Erie to Cleveland where you can take a stagecoach to Warringville. Always remember what Herr Goettel told us in school: think twice before trusting a stranger.[15]

The movement of such immigrants and European-manufactured goods on the Erie Canal epitomized the shift taking place in the nineteenth century. In the 1830s, just after the canal was completed, westbound traffic on the canal was six times that of eastbound traffic; people and goods were flowing to the West. By 1847, the West was becoming a tremendous source of goods; the tonnage of goods moving east exceeded that of goods moving west. By the start of the Civil War, eastbound freight was five times that of westbound freight. The hinterlands had become a source of excess, and that excess flowed through the Erie Canal.[16]

All of this growth along the Erie was to the detriment and dismay of the many other canal companies in other states. As difficulties mounted for canal and navigation companies elsewhere and traffic was concentrated toward the Erie, private capital for other canals began to dry up. Revenue along the Potomac exceeded expenses only in good years, and profit for investors was only 1.5 percent—hardly enough to lure additional capital.[17] As private capital continued to shrink for almost all canal and river companies, state governments constantly had to step in and increase their own already tremendous investments rather than letting the companies abandon half-completed projects. Canal companies, like the auto or banking industries of the early twenty-first century, had become too big to fail. Nationally, between 1820 and 1824, states invested (by going into debt) over $13 million in various projects and programs; by 1837 they had taken on another $108 million in debt. Over half of this public debt was for canal building alone.[18]

The national government remained isolated from this growing canal craze, even when things went badly. Starting in the mid-1830s, the canals went bust—and with them the states—in America's first great depression. The Panic of 1837 was driven by the enormous debt states had taken on to prop up their canal and river navigation companies; when the companies went bankrupt, they took state treasuries down with them. Most of the chastened state governments passed debt limit laws, self-constraining their ability to take on such massive, unsustainable debt in the future. This financial experience also reinforced the rationale of fiscal

federalism of the time, ensuring separation of the national treasury from those of the states.

New York was the exception. In 1837, while most other states were
spiraling toward bankruptcy, the Erie Canal commissioners reported
that the entire debt for the canal had been paid. The Erie Canal had
punctured the Appalachians and linked the Great Lakes with the
Atlantic in a way that no other Eastern Seaboard river could, and
for decades it remained the dominant path for commercial traffic. In
1905, the State of New York began work on the larger Barge Canal.
The canal was completed in 1918, although it was never as successful
as the original canal in luring waterborne traffic through the region.
The Barge Canal takes a slightly different route from the original,
bypassing some of the towns that the original canal helped create. It
even misses the town of Rome and the Erie Canal Village Museum,
which sits beside remnants of the original canal. Today the Interstate
highway and railroad lines run parallel to the Barge Canal with its
yachts, motorboats, and soon-to-arrive barges of oats from Canada.
Commercial traffic moved to rail in the late nineteenth century, then
to trucks in the twentieth; the state no longer modernizes the canal
but maintains its condition using Thruway tolls on trucks and cars.
Despite its diminished stature over time, the Erie Canal is partly
responsible for the shape of the contemporary United States. It positioned New York to be the commercial heart of the nation it is today,
and it established the Hudson as the riverine focal point for the first
half of the nation's existence and for its infant military.

Travelers going up the Hudson from New York Harbor toward
Albany and the Erie Canal will find a notably tight S-curve meandering around a point that juts out from the west bank of the river.
This place on the river has long been noted for its potential strategic importance. During the Revolutionary War, the Hudson River
was so easily navigated that the British navy sailed quite effortlessly
upstream to quell the rebellion in the Hudson Valley in 1777. To stop

the ships from moving up the river, local patriots pulled an enormous chain across the Hudson at this storied point. The British ultimately dismantled the chain and proceeded upstream to the then capital, Kingston, which they promptly burned to the ground. A year later, however, American rebels successfully defended the point. The rebels hung another chain across the river and then built a fort that eventually would guard an arsenal. The fledgling army later built a parade ground nearby; for the next two centuries, military cadets practiced marching there—at the U.S. Military Academy, also known simply as West Point.

Locating the U.S. Military Academy at a strategic point on a river was no accident. In the early days of the military academy, its primary focus was training future river engineers. At the turn of the nineteenth century, the United States was a technological backwater. Certainly there were individual scientific luminaries like Benjamin Franklin and Thomas Jefferson. But there were no sources of technical training. There were few engineers and even fewer institutional pathways to develop engineers. This lack of engineers had been a challenge for the Continental Army, as George Washington noted only seven days into his command. In April 1776, a recruiter was sent to Paris to ask for reinforcements, munitions, and good engineers. At least sixteen engineers came to support the cause. These French engineers did the planning and building that is critical yet often overlooked in long military campaigns; their efforts ranged from designing fortifications for the Continental Army to outlining the siege of Cornwallis at Yorktown. Among them was the young Pierre L'Enfant, who proved essential to the war effort and later would lay out the plan for the city of Washington, D.C.

At the end of the Revolutionary War, almost all the formally trained engineers were associated with the Army. But under the Constitution, it was not clear that there should be an army of any kind once the war had ended. The states were made up of former colonies that had found many of their complaints against the British bound up with the actions and behavior of its army during peacetime. Ever fearful of electing a potential tyrant as president, particularly when the president was a former general, those skeptical of a centralized national government sought to constrain such a tyrant's potential by restricting the country's ability to maintain an army to times of war only. Even if the army itself were not active in oppression, the high costs of sustaining an army would require taxes that could be their own form of oppression, or as James Madison expressed it in *Federalist* No. 41, "Liberty crushed everywhere between standing armies and perpetual taxes."[19]

Hamilton defended the need for a standing army and navy through his typical lens: commerce. As he put it, the entire purpose of uniting the states was "the common defense of the members; the preservation of the public peace, as well against internal convulsions as external attacks; the regulation of commerce with other nations and between the States."[20] These central functions of a government would be impossible to execute without a military that could be mobilized in peacetime, as well as during war, to ensure that the people of the nation would not be "subjected to continual plunder."[21] Besides that was "a constant necessity for keeping small garrisons on our Western frontier," into which settlers continued to move.[22] As to whether such security and defense of commerce justified the expense, Hamilton noted that the United States was encircled by other nations—Britain, Spain, and the Native American nations—and that some states, particularly New York, were more directly exposed than others. To support commerce between the states and avoid placing too heavy a burden on any particular state, the United States had to maintain a standing army to conduct military-related functions beyond fighting in battles.

Thus, though small, the national government did include a standing army and navy. Beyond maintaining garrisons or using a navy to protect against piracy, military duties included planning and construction of the fortifications that would be necessary in case of future wars. At the time, the young nation effectively had no engineers who could design and oversee such work. Although well equipped to train in the classics and law, the universities had only a smattering of actual scientists and no engineers. Thomas Jefferson saw the lack of engineering training as something to be directly addressed when he became president in 1801. Accordingly this president, so linked with the University of Virginia's pursuit of classic training in the liberal studies, established the first American university for the technical training of military officers and engineers.

For instructors, Jefferson looked overseas. The question was what nationality of instructors to hire for the academy. The British model of engineering was based on practical, on-the-ground training either with or without formal education. The French model was for students to attend an elite academy where they would learn the basic sciences and theory before moving on to actual engineering. Washington had relied on British engineers for advice on building canals along the Potomac. But Jefferson was a Francophile: he spoke French; he had lived in Paris as U.S. Secretary of State. And hostility still simmered between Britain and the United States. With these considerations, Jefferson adopted the French model in almost every way: he envisioned a military academy where an elite cadre of officers would be trained in the theoretical basis for the sciences and would then fill the intellectual ranks in the military hierarchy. Following the model suggested by L'Enfant immediately after the Revolutionary War, Jefferson developed a plan for the United States to have a department of public construction whose ranks were filled with academically trained engineers who would be charged with, among other things, "the means of raising or changing the course of water."[23]

Thus President Jefferson permanently established the Army Corps of Engineers, along with a military and engineering academy at West Point, in 1802. This academy became the nursery for U.S. military leaders and engineers, and it kept its strongly French roots; for two decades, four of the seven professors were French, and almost half of the textbooks used were available only in French. It was not until 1824 that another university focused on technical and engineering training—Rensselaer Polytechnic Institute—was founded, also along the Hudson River in New York.

The line between engineering training and military training at West Point was blurry, but the reputation of the army engineers was strong thanks to their rigorous training and their strong intellectual—if theoretical—grounding. For decades, the Army's elite first received engineering training at West Point on the banks of the Hudson. Only then did they begin their careers as river and harbor engineers, and eventually as military commanders. By the second half of the nineteenth century, the Corps of Engineers was so selective that only top graduates of the Military Academy were steered into the engineering corps pipeline. Robert E. Lee was a paragon of this system. Graduating at the top of his class at West Point in 1829, Lee chose to spend his required time in military service in the Corps of Engineers. He ultimately worked with the Corps for two decades, during which he designed and helped build coastal fortifications in Savannah and Baltimore and flood control levees along the Mississippi River at St. Louis. He later moved on to military posts at the Western frontier.[24]

Although training at the Military Academy at West Point has evolved significantly from its initial focus, West Point alumni continue to fill the upper echelons of the Corps of Engineers. When critical decisions are made about rivers in the United States, the person behind them is likely wearing an Army uniform and is often a graduate of the U.S. Military Academy. Part of this river management legacy derives from the Army's very early role as a source of engineering expertise, a role made possible because of the young nation's resolution to have a standing army. Another part of this Army-based legacy in river management stems from a particular Supreme Court decision that centered, again, on the Hudson River and profoundly changed America's technological and legal landscape.

Only a few years after Jefferson established the Corps of Engineers, a loud, smoky boat chugged by West Point at a speed slightly slower than the average walking pace. It was the first sign of a new era. In 1807, Robert Fulton had loaded up his first steamboat outside

New York City and headed for Albany. This first trip was a proof of concept. When he made it, everyone knew that river navigation had changed. When Fulton demonstrated that steamboats could move with or without wind, and with or without tide, he made towpaths irrelevant.

Fulton knew that what he had come up with was going to be worth big, big money. The State of New York did too. So did other states. All of them wanted Fulton's steamboats on their rivers, and so they competed to lure Fulton's company. But New York, ever the entrepreneurial state, won out by offering Fulton exclusive rights for constructing, operating, and employing steamboats on the waters of the entire state of New York—a statewide hydraulic monopoly.

As entrepreneurs, Fulton and his business partner Robert Livingston didn't want to do all the work themselves. They quickly created a franchise by leasing certain portions of their steamboat monopoly—specific river routes—to other operators. These other steamboat operators used the patented Fulton-Livingston steamboat technology on those routes and paid a small royalty to the founders. Among the takers: Aaron Ogden, who leased the route across the Hudson from Manhattan into New Jersey.

Steamboating was big business on the Hudson; it attracted go-getters and competition. Another entrepreneur, Thomas Gibbons, set up a steamboat line unconnected to the Fulton-Livingston monopoly and began directly competing with Ogden's line across the Hudson. To help with this new competitive line, Gibbons hired an 18-year-old unknown but ambitious ferryman named Cornelius Vanderbilt. Vanderbilt's entrepreneurial edge vaulted him to the role of business manager of the new steamboat service, which quickly drew traffic away from the original, state-sanctioned line. Ogden cried foul and sued, saying that the monopoly granted by the State of New York to Fulton-Livingston, which he leased, should preclude any competition from any other steamboats on the route. Gibbons said that the link to New Jersey across the Hudson made steamboating interstate commerce, and that a monopoly granted by New York didn't matter. The case went to court.

Decided in 1824, *Gibbons v. Ogden* would become a cornerstone case of American jurisprudence. It was heard by the Supreme Court under Chief Justice John Marshall, who was in the midst of establishing the role of the judicial branch out of the vagaries of the Constitution. This role of the judiciary in interpreting the Constitution for specific implementation, and the role that Marshall eventually adopted, is summarized in *The Federalist* No. 78:[25]

> The executive not only dispenses the honors but holds the sword of the community. The legislature not only commands the purse but prescribes the rules by which the duties and rights of every citizen are to be regulated. The judiciary, on the contrary, has no influence over either the sword or the purse; no direction either of the strength or of the wealth of the society, and can take no active resolution whatever. It may truly be said to have neither force nor will, but merely judgment; and must ultimately depend upon the aid of the executive arm even for the efficacy of its judgments.

The legislative branch had the clear role of establishing laws and appropriating money to carry out the functions of government. The executive branch had the authority to enact those laws. But the role of the judicial branch was less clear. At its simplest, the duty of the judicial branch was to determine when Congress passed acts that were "contrary to the manifest tenor of the Constitution." Judicial authority extended not just over rules developed by the U.S. Congress but also over those established by the state governments, when they conflicted with the Constitution or each other: "The judiciary authority of the Union is to extend . . . to controversies between two or more States; between a State and citizens of another State; between citizens of different States."[26]

In *Gibbons v. Ogden*, the controversy was whether one state could establish a monopoly in a way that affected interstate commerce. Given that the need for federal regulation of river commerce was the seed that sprouted into the Constitutional Convention, the Court found in favor of Gibbons. A monopoly on steamboat traffic between states was unconstitutional; New York couldn't tell New Jersey what was or was not legal on the interstate waterway of the Hudson River. Marshall read the commerce clause of the Constitution as giving Congress—the federal government—authority over interstate commerce; that ruling became the central contribution of *Gibbons v. Ogden* to jurisprudence. Based on Marshall's interpretation, "navigable waters" fell under federal authority, not the authority of the states. If you could get a boat on a river and move something commercial between states, then the federal government had authority over that piece of water. Or, as Marshall summarized it in the Court's decision, "The power over commerce, including navigation, was one of the primary objects for which the people of America adopted their government."[27]

Yet like hundreds of rulings to follow, even though the decision in *Gibbons v. Ogden* established the appropriate interpretation of the Constitution, the Court could not enforce that decision and would not lay out specific details of how it should be implemented. It was up to the executive branch to put judiciary branch decisions into practice; the legislative branch would then enact laws reflecting judicial interpretation. *Gibbons v. Ogden* reinforced the federal government's mandate to take charge of navigable rivers, and the only federal agency with some kind of role and expertise in navigable rivers in existence at that time was the U.S. Army Corps of Engineers. Thus the Corps of Engineers became and remained the primary federal agency tasked with river engineering and management, and the Corps eventually became the engineer in chief of rivers.

Also in 1824, Congress passed one bill for the clearing of sandbars blocking navigation on the Mississippi and Ohio Rivers—both commercially navigable rivers—and another bill for surveying routes for roads and canals "of national importance." In both instances, the executive branch tasked the Corps with responsibilities. Although the Erie Canal—an entirely intrastate waterway—was left to the State of New York, when Congress ordered the building of locks and dams along the interstate Ohio or Illinois Rivers, the Corps would do the work. Eventually the Corps became the point of contact for all riverine matters. Almost two centuries later, the Corps of Engineers oversees an immense array of river-related decisions, ranging from constructing hydropower dams on the Columbia River to managing the pseudo-riverine Everglades. When the Clean Water Act was passed in 1972, the agency charged with implementing significant portions of this new environmental protection regulation was the Corps of Engineers. This extension of constitutional interpretation likely would have stretched the imagination of Hamilton, Madison, and Jay; nevertheless, it springs from their ideas on how to interpret the Constitution as articulated in the *Federalist Papers*.

But *Gibbons v. Ogden* did something else that was just as important for the national economy, and eventually for the geography of the nation. By eliminating state-granted monopolies for river commerce, the Court's decision opened rivers to full competition between carriers—anyone with a steamboat, flatboat, or raft. Commercial river traffic for the next century and a half would be determined not by the central authority of a canal company or a state commission, but by the whims of hundreds of independent, freewheeling boat captains who plied the muddy waters of the midsection of the continent.

CHAPTER 2

Life on the Mississippi

"Wornicott Bay. Destruction. Glassock. Fairview. Black Hills. Jackson Point."

Donnie Randleman had just climbed the four floors of steep stairs into the wheelhouse of the towboat *Christopher Parsonage*. He was a bit short of breath as he spouted off these random names in response to my questions about where the boat was at that moment and what points of reference would be coming along next. Randleman is a barrel-chested, grumpy twenty-first-century towboat pilot who looks like he might be more at home at a Harley Davidson rally than in a towboat wheelhouse. He is coming up to the wheelhouse to take a turn piloting the *Parsonage* and relieve the boat's captain, who has been on since 6 p.m. The captain is Robert Duty, known up and down the Mississippi River as "Howdy Duty." He looks—and chuckles—like a young version of Santa Claus. This particular shift change occurs at midnight during an ink-black, moonless night on the lower Mississippi River, somewhere in a desolate region of northern Mississippi–Louisiana.

Howdy is finishing up his six-hour shift of piloting amid a glow of radar screens, GPS screens, and sonar readings that give him a steady stream of data: a twenty-first-century view of the Mississippi

River. In addition to piloting a towboat, Howdy is a perpetual source of stories—which he strings together one after another. Each bend in the river or each pilot that passes by reminds him of another story, most of them having to do with an event that took place on the river. And he says some of them may even be true. When Randleman arrives at the wheelhouse, waiting to take over steering the barge tow down the river, Howdy's attention is focused intently on a narrow cylinder of light emanating from the spotlight on the roof of the wheelhouse. The light now focuses out across the river, lighting up a circular area of the riverbank that is 15 feet in diameter and a quarter mile away. Howdy occasionally swings the spotlight over to illuminate one of the passing red buoys that mark the river's main navigation channel—a narrow ribbon of water that the pilots can trust is kept sufficiently deep for their boat. After spotting the buoy and adjusting his path slightly, Howdy swings the cylinder of light back to the riverbank to show Randleman where they are. From this brief, dark glimpse of riparian greenery on a moonless night, Randleman can pinpoint the exact location of the *Parsonage* within an eighth of a mile and name every upcoming bend, stop, bar, and point in their correct order—all this at midnight, before having any coffee.

This view of the Mississippi River—looking from the river out, rather than from the land in—is rare in the twenty-first century. The lower Mississippi River, from Saint Louis to New Orleans, is now an industrial superhighway. Taking a small ski boat, or even a fishing boat, on this stretch of river would be like riding a bicycle amid the trucks and trailers of the New Jersey turnpike. Howdy and Donnie's task on this particular run of the *Parsonage* is to move thirty barges to New Orleans. A barge is simply a floating container, each one 35 feet wide, 200 feet long, and up to 12 or 13 feet deep; when loaded, a barge weighs 2,000 tons. The crew straps together the barges into a collective raft, referred to as the *tow*. When considered as one unit, this tow being pushed by the *Parsonage* is 210 feet wide and 1,400 feet long, in addition to the 180-foot-long towboat itself. A barge tow of this size is large enough to hold two *Titanics* side by side on its top, with plenty of room to spare.

In its heyday of the mid-nineteenth century, a full boat on the Erie Canal could move a couple hundred tons; each individual barge now being pushed by the *Parsonage* is ten times larger than a boat on the Erie Canal. When considered as a whole, the total weight of the thirty-barge tow pushed by modern towboats like the *Parsonage* can exceed 60,000 tons. To appreciate just how much this is, compare it to two other modern modes of heavy transport: trains and trucks. A single barge carries the same amount of weight as sixteen fully loaded rail cars or seventy tractor-trailers. It would take over four hundred rail cars (requiring a dozen locomotives) or more than two thousand tractor-trailers to move the same amount as Howdy Duty's boat is moving in a single tow.

Typically, boats get their maneuverability from their depth in a river, which allows their hull to work like a massive rudder. But riverboats on the Mississippi spread out the weight of the cargo over an enormous area, which allows them to carry more weight in shallow water. Piloting a barge tow is less like driving a car than flying a kite. Barge tows skim across the top of the water rather than slicing through it; and all the while, the current is moving beneath them in multiple directions. Pilots like Donnie and Howdy don't steer the quarter-mile of barges around sweeping meander bends so much as they slide them. They have to negotiate hundreds of meander bends perfectly so that when they come out of one bend, they are positioned just right to get around the next, or to slip their barge tow through the narrow span between the bridge piers at Memphis, Greenville, or Vicksburg. They have to do all of this whether the river is flowing very low or boiling and roiling at flood stage. And they have to do all this at night as well as day. This is the reason Donnie can rattle off the names of all the upcoming waypoints along the river in the middle of the night. He must always be keeping a mental checklist of what's coming and in what order, so that he will know how to orient the trajectory of this unwieldy craft while negotiating the upcoming sequences of turns and obstacles. Donnie, Howdy, and every other boat pilot have memorized the river.

The construction of modern barge tows itself is as impressive as the skill of the boat's pilots. This particular run of the *Christopher Parsonage* is a "milk run," picking up and dropping off individual barges along the river in addition to delivering the main tow in New Orleans. This means that the two pilots have to plan on dropping off two barges in Memphis, another in Rosedale, Arkansas, and then pick up two more in Greenville, Mississippi. The process started in Cairo, Illinois, where the deck crew began by "building tow," the physically grueling process of taking dozens of unruly barges that were being pushed by smaller boats—amid the river's currents—and assembling them into a single tow that the *Parsonage* can manage.

The deck crew of two or three men per shift jump on and off arriving barges, strapping and cabling the newly arriving barges onto the growing tow—a massive raft held together with a series of 50-foot long, inch-thick steel cables, each weighing nearly one hundred pounds. To these cables are added immense steel ratchets, chains, and hooks that give the tow deck a rusty medieval look. Building tow in the summer involves six-hour shifts of steadily lifting awkwardly shaped weights on top of a massive steel skillet that simmers and shifts in the southern heat. Howdy Duty and Donnie Randleman, like all captains, began their careers by working for years as deck hands, in Howdy's case starting at age 18. Now, as a pilot and captain, Howdy looks down from the air-conditioned wheelhouse four stories above at the deck hands building tow and chuckles, "Yeah, it's like prison with a chance of drowning."

With the tow assembled, the *Parsonage* slips its quarter-mile-long tow into the main current of the Mississippi and heads south, away from a remarkable riverine juncture: just upstream are the Ohio and Tennessee Rivers to the east, the upper Mississippi and Illinois Rivers to the north, and the Missouri to the west. Up the Mississippi and Illinois to corn, or up the Missouri to the wheat of Nebraska and the Dakotas, or up the Ohio to Appalachian coal and Pittsburgh steel. Navigating and commerce on the Mississippi River has tied together distal economies and seemingly irreconcilable societies: a century and a half ago, going up the river led to freedom while going down the river led to slavery. Even today, the Mississippi still weaves together disparate regions. Pilots spending the days and nights along the river listen to brief radio conversations in strange accents from other boat captains: flat nasals from Wisconsin, southern drawls from Alabama, twangs from Arkansas, and the impenetrable Cajun patois from Louisiana. All these captains push all these barge tows up and down a river buzzing with traffic that is moving the heaviest, bulkiest goods of twenty-first-century America.

The Mississippi River of the early nineteenth century was, for the United States, only beginning to be appreciated and accessible. The infant United States had gained the trans-Appalachian region—extending from the western edge of the original thirteen states to the Mississippi River—with the Treaty of Paris that ended the Revolutionary War. But America had to share access to the Mississippi for years. New Orleans remained first in Spanish control, then in French control until 1803, when Jefferson acquired 828,000 square miles from France in the Louisiana Purchase. This territorial expansion not only doubled the size of the United States but also added the tremendously important port of New Orleans. With New Orleans and the Missouri River basin, combined with the rest of the Mississippi River basin that America held already, large and navigable rivers became central features of America's geography.

These newly acquired rivers were remarkably different from those of the Atlantic coast around which the nation had formed and grown. They were especially different from the economy that had developed around the Erie Canal, where settlers and goods were concentrated along the lone commercial highway managed and operated by the New York state government. In stark comparison, the network of the Mississippi–Missouri, Mississippi–Illinois, and Mississippi–

Ohio—along with countless others—was far-reaching and dendritic. The Mississippi and its associated rivers diffused settlers along their alluvial veins and into the farthest reaches of the hinterlands. On the edge of the frontier, settlers harvested wheat, coal, and cotton that were then passed along the riverine arteries toward the heart of the alluvial empire—New Orleans. This quixotic sub-sea-level city grew faster than any other American city from 1810 to 1820, when it became securely established as the country's second-largest exporting city, behind only New York.[1]

By the middle of the nineteenth century, steamboats could easily pass up the hydraulic superhighway of the Lower Mississippi from New Orleans to St. Louis. Without the assistance of locks or dams, some boats then continued over 2,200 miles up the Missouri River's winding course from St. Louis all the way to Fort Benton, Montana, where the river borders the emerging wheat fields of the Great Plains. Other steamboats plied the seemingly endless rivers, swamps, and sloughs of the Deep South for ever-lucrative cotton. Traffic along the Erie would cease for months while the canal was frozen over in winter, but New Orleans persistently teemed with business—suffering only the occasional light frost. The Mississippi River basin was the brave new world of America's riverine landscape.

The changing technology of boats contributed to this new economy. Companies approved rapid and constant adaptations in the design of steamboats to increase their carrying capacity while making their draft shallower. The new designs allowed boats to carry heavier loads through what had previously been impenetrable tributaries, swamps, and sloughs. In the 1840s, a moderately sized 170-ton steamboat carrying 50 tons of freight and 80 passengers—already larger than those boats then operating on the narrow Erie Canal—could operate in only 22 inches of water. Steamboat pilots, well known to be an arrogant and competitive lot, boasted that all they needed was "a heavy dew" to run their boats. Between 1815 and 1860, the economic productivity of steamboats advanced at a rate exceeding that of any other method of transportation at the time, including railroads.[2]

Steamboat technology combined with the physical characteristics of southern and western rivers to create a distinctive political economy in the region: there was little reason for the government to get involved, and thus farmers and steamboat pilots were comparatively free from government tolls or intervention. When geographic barriers like the Fall Line were overcome by canal and a government-backed lock and dam, tolls were set by a central authority like a state canal commission, as much by politics as by necessity. Planning and engineering were predominant in ensuring that canal-based transportation remained open and efficient. On large rivers, however, commerce was moved by powerful steamboats that could go about their business while largely immune to the vagaries of politics and government funding. When politics and funding did enter the equation—for dredging sandbars or channelizing a particular meander bend—the goal was to ensure clear commercial navigation on interstate rivers, thus allowing the rivers to remain toll-free. Steamboat pilots could be very independent because the rivers of the Deep South and Midwest were massive in comparison to their eastern or northern counterparts. The Erie Canal was 4 feet deep and 40 feet wide when it opened in 1825, and it could handle boats up to 30 tons; when initially expanded in 1862, it was 7 feet deep and boats up to 240 tons could pass through. In 1860 a steamboat plying the Mississippi could carry over six hundred tons of freight, excluding passengers, and operated independently of anything other than occasional docks needed for loading and unloading. The Mississippi and Missouri were often thousands of feet wide, and their primary tributaries—the Ohio, Tennessee, Illinois, Iowa—were not much smaller. The skill involved in piloting western rivers was adjusting the path of the boat rather than the path of the river. Engineers who designed canals were respected, but steamboat pilots on the Mississippi were revered. Any pilot who could steer a steamboat, or any entrepreneur with a flatboat, had the ability to move among thousands of miles of waterways to make their fortune.[3]

And make their fortune they did. They were part of a profession that was as hazardous as it was lucrative. The average life span of a steamboat in the mid-nineteenth century was typically only a few years; the combination of shoals, logs, ice, and adolescent steam power technology resulted in a constant turnover of boats and cargo. But risk and return are tightly entwined. On the particularly nasty Missouri River, it could take less than two round trips for an entrepreneur to pay off the cost of building a steamboat, so a pilot's monthly take-home revenue was often as high as 5 to 10 percent of the total value of the boat. And a century and a half later, Howdy Duty noted that because he didn't get a college degree, piloting a barge tow was one of the few careers that paid enough for him to afford good health care, for his wife to stay home and raise kids, and for helping their kids go to college—assuming they don't become riverboat pilots like him or his dad.[4]

Likewise, any riverfront farm or plantation with a dock could count on an endless flow of steamboats passing by, all clamoring for the business of moving the goods upstream or downstream. In the alluvial empire of the Mississippi, the only investments needed were enough capital for a single steamboat for the movers and enough capital for a river dock for the buyers and sellers. The size of the rivers and the number of riverfront farms created intense competition, eliminating any potential monopoly of river commerce by a lone entrepreneur or company. And thanks to John Marshall and the Supreme Court, state governments were prohibited from sanctioning monopolies for routes along the rivers. Prices for moving goods along the rivers of the South and West were kept low by competition rather than by a canal company or a state legislature setting tolls.

America's distinct riverine regions were differentiated not only by their technologies and economies but also by their art and literature. The art of the Hudson River School in the mid-nineteenth century was characterized by misty, pastoral, riverine landscape scenery;

people appeared in corners of the paintings only to give a sense of scale or even a sense of the region's wealth. But the art of the Mississippi region during that time was of the people, such as the paintings of flatboatmen by George Bingham. The diverse stories, phrases, and caricatures of the newly settled mid-continent were also constantly being funneled and amplified along the Mississippi River, putting young "cub" steamboat pilot Samuel Clemens in a great position to observe contemporary society.

Pilots like Clemens constantly communicated with people along the river while making their numerous stops to pick up and drop off passengers or cargo. But like Howdy Duty and Donnie Randleman 150 years later, those early pilots also learned the river and could read the boiling currents to steer clear of underwater shoals. Rather than using GPS and underwater sonar to measure depth, pilots like Clemens depended on leadsmen who poked and prodded the river with their staffs to measure depth. These leadsmen would read the river's depth in fathoms—one fathom was called a mark. A safe depth for a steamboat, and one that a cub pilot wanted to hear the leadsmen yell out, would be two fathoms, or in the slang of 1850s rivermen, "mark twain"—an apt pen name for Samuel Clemens, an author who began his career as a steamboat pilot. Given the novelty, rambunctiousness, and enthusiastic optimism of the region, it's no surprise that the chief satirist of the nineteenth century grew up on the banks of the Mississippi.

In the 1880s, 25 years after the real heyday of Mississippi steamboats, Mark Twain returned to the Mississippi to see the river again—and found that it had been reshaped by government, technology, and the economy. Thomas Jefferson's Corps had begun as a slim collection of Army officers with a nebulous mission. In the *Federalist Papers*, Hamilton argued that the national government should be able to use the taxation mechanisms available—levies on imports, land taxes, or poll taxes[5]—to the degree necessary to support the functions that the national government had adopted: "A government ought to contain in itself every power requisite to the full accomplishment of the objects committed to its care, and to the complete execution of the trusts for which it is responsible, free from every other control but a regard to the public good and to the sense of the people."[6] Initially, the national government relied solely on taxing imports because its role was so restricted that its expenses were quite limited. Up to the 1820s, the national government had no role in river navigation.

But in 1824, the year the Erie Canal was being completed and a year after John Marshall declared the federal government the regulator in chief of interstate commerce, Congress approved a seemingly innocuous bill: "An act to improve the navigation of the Ohio and Mississippi Rivers" was the first act that inserted the federal government actively into river navigation. This bill was targeted carefully by constraining the federal government's actions to rivers that were viewed as essential for interstate commerce. The Corps of Engineers put this legislation into action.

Following this initial toehold came a string of congressional bills, one passed every few years. Each bill gradually expanded the functions and geographic scope of the federal government over rivers, and each one was executed through the Corps of Engineers' Mississippi River Commission.[7] Twain himself was doubtful, even scornful of their ability to harness the river:

> One who knows the Mississippi will promptly aver—not aloud, but to himself—that ten thousand River Commissions, with the mines of the world at their back, cannot tame that lawless stream, cannot curb it or confine it, cannot say to it, Go here, or Go there, and make it obey; cannot save a shore which it has sentenced; cannot bar its path with an obstruction which it will not tear down, dance over, and laugh at. But a discreet man will not put these things into spoken words; for the West Point engineers have not their superiors anywhere; they know all that can be known of

their abstruse science; and so, since they conceive that they can fetter and handcuff that river and boss him, it is but wisdom for the unscientific man to keep still, lie low, and wait till they do it.[8]

Despite Twain's doubts, the Corps perpetuated its "abstruse science" to great effect. At first the Corps of Engineers followed the restraint of federal government that was envisioned in the *Federalist Papers* and had been typical of the eighteenth and nineteenth centuries: the agency dredged an occasional bar, or straightened a meander bend here and there, but overall its work was sporadic. All the way through and after the Civil War, the federal government remained hesitant to interfere, always questioning whether a particular river project was genuinely a federal interstate issue or whether it should be left to the states.

But during the Great Depression, the nation discovered that river projects were particularly useful for absorbing labor. Any semblance of fiscal or federal government restraint, considered so essential in the era when the *Federalist Papers* were written, was largely abandoned. Franklin Roosevelt unleashed vast federal funds—or at least vast federal debt—in service of river navigation. Until then the federal government had gone into debt only in times of war. During the first national economic depression, triggered by the canal collapse in 1837, the federal government did nothing to bail out the economies of state governments. But a century later, during the Great Depression, the federal government inserted itself as the centerpiece of fiscal activity while states became relatively inert. Thanks to this paradigmatic shift, federal spending on rivers rose from $57 million in 1929 to over $178 million in 1937.[9] River navigation projects were a preferred spending option, in part because they had long-term economic benefit and absorbed labor. The construction of Fort Peck Dam in remote Montana alone employed over 11,000 people and began a long process of harnessing the Missouri River for increased river traffic. On the Upper Mississippi River, decades of previous attempts to increase traffic had largely failed. Nevertheless, the New Deal era pumped enough funding and labor into projects that by 1940, the Upper Mississippi River had twenty-six locks and dams providing a continuous 9-foot-deep navigation channel between St. Paul and St. Louis.[10]

At the center of this flurry of river building was the Corps of Engineers, whose hallmark was straightening and deepening—channelizing—rivers. Meander bends made the path for boats and barges tortuously long and increased the prevalence of hazardous sandbars and mixing currents. To increase the ease of navigation and decrease unemployment rates, channelizing rivers became a favorite project for Corps bulldozers and dredges. The agency focused much of its uniformity-loving efforts on the Lower Mississippi, which was shortened by over 150 miles; the stretch of river passing Greenville, Mississippi, alone was reduced in length from 51 to 24 miles. Nationwide, from 1936 to 1972, the Corps channelized over 11,000 miles of rivers, dramatically reshaping the riverine landscape.

The Corps didn't stop with channelizing, and it didn't stop when the Great Depression ended. In 1948, the Corps of Engineers was home to 200 army engineers, 9,000 civilian engineers, and 41,000 civilian employees. Half a century later, at the turn of the twenty-first century, the Corps was overseeing $62 billion worth of civil works projects and had contracted only slightly, to a workforce of 35,000—still employing more Americans than the Departments of Energy, Labor, and Education put together. And the Corps was becoming increasingly synonymous with pork barrel projects; for Corps projects authorized in 1986, at the end of the Reagan administration when funding was comparatively restrained, the greatest determinant in getting a Corps-related project funded was whether the project was located in the home state or district of a House or Senate leader. Two decades later, when the first of the twelve bills George W. Bush vetoed as president concerned funding for the Corps, both the House and Senate overrode the veto handily. They were all keen to keep federal funds flowing to their states via Corps projects.[11]

This history of Corps projects, and even the pork barrel elements, is not lost on Howdy Duty or Donnie Randleman. Yet each river bend removed and each upstream dam constructed to moderate downstream flows makes their task of navigating the river that much easier. Howdy has a habit of pontificating at night while piloting the *Parsonage*, and politics and river history are some of his favorite subjects. The themes of his monologues vary tremendously, including treatises on the efficiency of barging commodities compared to trucking them, the centrality of commercial traffic for the agricultural sector of the economy, and the political history of the United States. For his own industry, he captures a complex history concisely and accurately, if not blithely. "Congress decided a long time ago that rivers were for commerce, and that's just the way it is. And they told the Corps to keep some of them rivers navigable for traffic like ours. And that's the way it is. And the Corps has done a fine job on this here river. They done made this here into a fine, easy goin' river."

The Corps had indeed worked its hydraulic magic on the Lower Mississippi. By the twenty-first century, the Corps had mostly tamed but not conquered the Mississippi. This level of effort means the *Christopher Parsonage*, like other towboats, has to stop only rarely. Howdy Duty and Donnie Randleman run the river 24 hours a day, 7 days a week, 365 days a year. As the crew often jokes, the only thing that would stop the *Parsonage* would be if they ran out of coffee or cigarettes. Towboat pilots spend their never-ending days and nights plying the river between the ubiquitous buoys—laid out and maintained by the U.S. Coast Guard—that mark underwater hazards, whether they are shoaling bars or the edges of infrastructure put in place by the Corps to "train" the river to be more navigable. Though largely hidden from sight, the basic federal infrastructure to keep the river navigable—rock weirs, bank protection, buoys, mile markers, dredged sandbars—is nonetheless expansive. Far upstream, in the remote reaches of the rivers that feed into the Mississippi, the Corps had decades earlier built massive reservoirs to store spring runoff and let it slowly leak out through the summer. To prevent the river from drying up in the summer, these dams release precisely calculated flows to ensure sufficient water depth for barge tows thousands of miles downstream, including those being pushed by the *Parsonage*.

Howdy doesn't need to know how the weirs work or how much water the various reservoirs store. The barges pushed along by the *Parsonage* require 12 feet of draft, and Howdy knows that the Corps of Engineers is doing whatever it does to constantly maintain a minimum depth between the buoys. He knows that the Coast Guard is regularly measuring the river depth and moving the buoys to mark the edge of the navigable channel. The federal government is responsible not just for marking twain, but for making it less necessary for towboat pilots to do so.

Even if it goes largely unnoticed, even if it's nearly invisible on the East Coast, with its nearly evaporated remains of the Erie Canal, river traffic is central to the national economy. Costs to move cargo by barge are shockingly low: In 2014, using the barge line of Howdy and Donnie to move a ton of grain from Minnesota to New Orleans cost just over six dollars. From Cairo, Illinois, to New Orleans, the cost is just over three dollars. River traffic on the primary stretches of the national waterway system has continued to increase: from 1924 to 1970, total river freight traffic grew from 34 million tons to 472 million; and by 2000, 880 million tons of freight moved over the nation's 25,000 miles of navigable waterways. And this is not just grain moving from the Midwest to New Orleans; just as much traffic on the Lower Mississippi is carrying petroleum products. As Howdy said in one of his evening monologues aboard the *Christopher Parsonage*, "You can't afford to move grain from the Midwest to New Orleans by truck, and barely by train. You gotta have rivers. And you gotta have barges."[12]

Modern commercial traffic on a large river can best be seen in Natchez, Mississippi, where the character of the Mississippi changes noticeably. All along their southbound run down the Mississippi,

Howdy Duty or Donnie Randleman radio to occasional oncoming towboats, coordinating how they will pass by each other. But the radio traffic aboard the *Christopher Parsonage* picks up in Natchez, becoming a constant background of conversation rather than an occasional call. The pilots start hearing foreign accents on the radio, and some of the first ships appear—vessels clearly not built for rivers. Along the river at Natchez, Corps of Engineers dredging barges are moored next to Coast Guard boats that are managing some of the flurry of activity. Howdy Duty becomes much less conversational in Natchez, instead focusing his attention on the constant cackle of radio negotiations to move the *Parsonage* around enormous oceangoing ships slowly winding their way along this deeply dredged portion of the river. By the time the *Parsonage* reaches New Orleans, the river is a jumbled mixture of barges, towboats, ships, grain elevators, refineries, and pipelines. Imagine a truck stop, a shopping mall parking lot, an industrial park, and a rest area all mixed together on the busiest travel day of the year; then picture everything inordinately larger and constantly in motion amid river currents, and you have an idea what New Orleans looks like from the Mississippi River.

Here, between Natchez and New Orleans, the past and present of America's rivers started to come into focus alongside each other. As the deck crew began breaking up the *Parsonage*'s tow, and smaller tugs swarmed around its shrinking carcass to take the different barges upstream and downstream to various terminals or to waiting ships bound for China, Howdy Duty was already getting his paperwork in order for the trip north. Just downstream was a massive tow of empty barges, all ready and waiting to be pushed back upstream to be refilled with the products of America's interior and then brought back to New Orleans for export.

In the 1770s, when the framers of the Constitution were interpreting its meaning and developing how it would be put into practice, ports along the Eastern Seaboard were the hubs of commerce. These rivers and ports—from Richmond on the James to Albany on the Hudson—played an essential economic role by sitting at the nexus of ocean and land and of eighteenth-century import and export. Down the rivers flowed rafts of timber or flatboats loaded with agricultural commodities of the growing farms and plantations; up the rivers moved ships carrying manufactured goods from Europe.

Howdy Duty's mixing of barges and ships in New Orleans is likewise the mixture of land and ocean—the twenty-first-century equivalent of mixing colonial era river flatboats and ocean schooners. The Mississippi River in New Orleans is, as envisioned in *Federalist* No. 2, an integrator of America. The mixing of drawls and twangs occurs because the work of moving commerce draws together citizens in a way that is inherently necessary. It is a starting point for binding separate peoples into an imperfect, yet potentially coherent nation.

CHAPTER 3

The Rise of the Levees

You can drive the two hundred miles of the Mississippi delta from Vicksburg, Mississippi, to Memphis, Tennessee, in two ways. Highway 3 skirts the eastern edge of the delta, riding along the edge through kudzu-covered, silty-clayey rolling hills. To the west, Highway 1 dives straight into the flatness of the delta. There are no rises and falls along Highway 1 and no topography in the seventy-mile-wide delta save for a single hill running parallel to the Mississippi River. This is the levee—the only geographic feature standing between the Mississippi River and 7,000 square miles of perfectly flat, fertile farmland.

Levees are the blunt instrument of flood control. In its simplest form, a levee is just a small mound of earth piled up along a river, providing an extra foot or two of elevation that may be just enough to keep the river confined to its channel during a flood. But at their grandest, levees are carefully planned, engineered, and managed infrastructures on which entire cities, regions, and economies depend for survival. In the Mississippi delta, miles upon miles of levees run along every major and minor tributary, swamp, and slough.

The scale and uniformity of the levee along the main channel of the Mississippi River is astonishing: a solid wall of earth with a base spanning over 300 feet, stretching almost four stories tall at the crest, with a wide road running along the top. Imagine filling a college football stadium with a trapezoid of compacted dirt up to the top of the bleachers, building a road on top, and then lining up those dirt-filled stadiums side by side, continuously, from Vicksburg to Memphis on both sides of the river. That is what it takes to put the Mississippi River into its hydraulic straitjacket. And the levee goes on south to New Orleans and north far beyond Memphis, stretching thousands of miles upstream to Minneapolis, as well as along every major tributary from other rivers like the Ohio and Missouri to every other stream that touches the river.

In May 2013, the Mississippi is just cresting at 1.26 million cubic feet per second (cfs, the standard descriptor of river flow, or discharge). To get a sense of just how much water goes down the Mississippi River, 1.26 million cfs amounts to 8.8 million gallons of water per second—enough to fill Lake Mead to the top of Hoover Dam in just over three days. This flood crest in May is actually the first of two pulses of water that will pass Vicksburg for the year due to the sheer size and diverse geography of the Mississippi River basin. This first pulse comes from the Ohio River valley and Upper Midwest, fed by all the thunderstorms and cold fronts that typically generate rain in the early spring. The second pulse is the snowmelt runoff from the Rockies and Great Plains, which works its way all the way through the gullet of the continent that is the Missouri River and then moves on through the Mississippi. The timing of the pulses is fairly regular, but the size can vary tremendously from year to year. The year 2013 is typical; the river is coming up to the top of its banks and at places reaches the bottom of the levee, but it is nowhere near the top of the levee, which can handle more than twice the flow of 2013.

Kent Parrish manages 410 miles of levees for the Vicksburg District of the U.S. Army Corps of Engineers. Because the river in 2013 is passing a low flow, Parrish is very calm, and so were others in the offices of the Corps of Engineers in Vicksburg. The Corps is the fed-

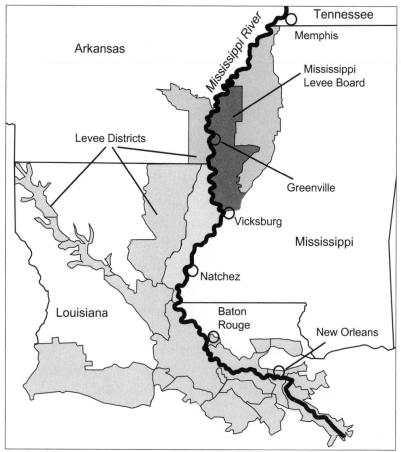

Major levee districts of the lower Mississippi River Valley, including the Mississippi Levee District and Greenville, fitted together like patchwork.

eral agency tasked with controlling floods. Its engineers build levees and dams, and anytime there is a flood, levee break, or some other hydrologic disaster on a river, the Corps is in the thick of things using military trucks and helicopters for a variety of flood-fighting functions. But the highly visible role of the Corps belies what actually happens before, during, and after floods in the United States.

Parrish, as one of the Corps members tasked with levee management, says that much of the involvement of the Corps during floods is mundane: "We at the Corps just *build* the levees." Parrish works largely out of the limelight, doing things like checking flood levels and making phone calls. The real work, Parrish says, is done by organizations in charge of the levees—the levee districts. They're the ones who *maintain* the levees and bear the brunt during the flood itself. Levee districts are obscure compared to the Corps, yet they are ubiquitous—that becomes apparent when you take a close look at a map of flood control infrastructure along a large river. Some levees are labeled as belonging to the Corps; other levees are labeled as belonging to a specific district. Many of these districts sit downstream of an enormous flood control dam built and operated by the Corps of Engineers. A map of flood control infrastructure is the best way to begin appreciating the layered nature of flood control. A brief look at a map of levees clearly shows that the Corps may not be as central to understanding flood control as these largely unheard of and ignored levee districts.

A levee district that Parrish works closely with is the Mississippi Levee Board, based about 100 miles upstream in Greenville, Mississippi. Greenville sits in the smack-dab middle of the Mississippi delta. Most cities along the Mississippi—Vicksburg, Yazoo City, and Baton Rouge, for example—developed in places where the river has meandered over to the edge of the valley and close to the bluffs. A city on a bluff has access to a river but protection from a flood. Greenville is close to the river but far away from either bluff. It is a city dependent on levees.

Downtown Greenville is nestled just on the landward side of the levee with its old downtown area starting at the toe of the levee. On the river side, the levee is paved to allow parking for a city park. Going up the side of the levee is a series of brass numbers and lines. Each line marks the elevation the river reached during a flood, and the brass numbers record the year. As the sun sets in the west, the shadow cast by the levee creeps from the streets and rises slowly up the sides of buildings in a sequence not unlike how water would rise in the city if the levee failed.

These levees are the literal and figurative turf of the Mississippi Levee Board, whose headquarters are small and inconspicuous: an

Makeshift sidewalks in Greenville, Mississippi, during the flood of 1927.

office building outfitted with cubicles and fluorescent lighting. But inside the building is a hallway lined with black-and-white photographs of former chief engineers going all the way back to 1865, when levees throughout the South were in tatters after years of war punctuated by three enormous floods. The people in these pictures— General S. W. Ferguson, William Starling, and William Elam, among many others—are well known in the field of river science and hydraulic engineering, the authors of hallmark scientific publications and books on all manner of topics related to flood control. Rural Mississippi, and specifically Greenville, does not seem like a natural hotbed for luminaries of fluid mechanics and hydraulic engineering. It might seem that such engineers would make their homes down in Vicksburg or New Orleans with the Corps. But the levee district here in Greenville has always been a hub of engineering—a legacy being upheld by the current chief of engineers, Peter Nimrod.

When he walks past the photos of his predecessors, Nimrod just sips from his coffee cup and slowly drawls, "Tha's raight, it's an intimidating hallway." Along with its hall of fame, the little office building has a conference room of sorts that also serves as an impressive library and museum of all things flood related. The walls are lined with photos of historic floods when things didn't go so well for the levee district: photos of refugee camps perched atop narrow levees surrounded by miles upon miles of water. Photos of the town of Greenville under five feet of water. Photos of sandbagging crews trying to fight off the last few inches of rising floodwaters. Photos that remind viewers of what the levee board works year round to avoid.

Nimrod and his assistant engineer Bobby Thompson take an hour to describe what their lives on the front line of flood control are like. They describe the levees they manage, from the massive mainline levee to the more obscure levees of the backwaters and tributaries. They know every nook and cranny of every mole hill and mobile home along the hundreds of miles of levees in their district. While Kent Parrish and the Corps downstream are in charge of the planning and the broader system of levees all along the Mississippi River, Nimrod and the levee board handle the weedy details of day-to-day levee management in their particular geographic area. One such detail is hiring people to constantly mow the grass on the levees. Another is the need for widening and thickening the levees, which leads to the chronic issue of dealing with landowners. The board is currently negotiating with the recalcitrant owner of a mobile home, whose residence is in the way of one of the areas they need to widen. This, even though the levee board has eminent domain rights and the levee is the only thing standing between the owner's home and the river. These are the kinds of things Nimrod and Thompson do in years like 2013, when the flow of the river is not too large. During floods, the levee district shifts into flood-fighting mode; it is the government entity inspecting every foot of levee night after night during floods, finding and organizing people to fill sandbags, and evacuating people from flood-prone lands in case the levee doesn't hold.

This seemingly haphazard division of responsibilities creates confusion about flood control. Who is in charge? Who does what?

Both the Corps and the Mississippi Levee Board are creations of government for flood control. Both have engineers, hydrologists, bulldozers, and boats. Both have different roles and responsibilities for flood control, but they overlap and intersect. This intersection, and the resultant confusion over who is in charge, is a product of federalism—the most essential element of American governance. In 1787, when the framers of the Constitution set to work on rethinking the structure of government, their foremost challenge was balancing the need for a more centralized national government against persistent desires for state and local autonomy. The result—federalism— is the notion that there should not be a single government, but rather what Woodrow Wilson would later describe as a "series of governments within governments."[1]

Under this system of federalism, the idea is that power should be shared and separated not just between branches of government but also between layers of government. From the federal government to the states and on down to counties, municipalities, villages, and townships, each layer would take on its own particular roles and responsibilities. The framers envisioned a national, central government that would provide for the broader interests only it could furnish, such as national security, economic development via free trade, or even basic scientific research. All other functions would be handed down to the lowest level of government—the closest to the community— that could handle the problem, since governments at this level best understand the true demands of their people and their region.[2]

That was the idea. To put the idea into practice meant that government functions would first have to be distilled through the ideals of federalism. Thus, for flood control, the appropriate question was not "Which *agency* of government should be responsible for flood control?" Rather, the central question was "Which *level* of government should have the responsibility for flood control?" The answer to that question, like almost all innovations and developments in flood control, came out of the Lower Mississippi valley.

Settlers who arrived in the Mississippi valley in the early nineteenth century were confronted with tens of thousands of miles of navigable rivers, bordered by some of the most fertile soils in the world. The pioneers' dreams of turning this landscape into an alluvial empire were limited only by flooding. Major cities of the Mississippi valley—New Orleans, Vicksburg, Memphis, and St. Louis—all sit on topographic oddities that keep them just out of the reach of floods. To turn the potential of the river valley into reality, however, farmers had to control floods—and that meant building levees.

At that time, the federal government was largely inert. The states took on almost every government role and responsibility, in turn seeking to push them on to local governments. Levee districts, initially formed in the 1830s in Mississippi and Louisiana, were the natural solution to flood control in this political climate. The entire state of Mississippi, for example, did not need to be taxed to accrue the funds to build levees, because much of the state was not flood prone. Instead, the Mississippi Levee Board, along with other districts along the river, was organized by planters in the flood-prone delta and recognized by the State of Mississippi as a formal unit of government, comparable to a municipal government of Greenville or Jackson. But whereas municipal governments have broad responsibilities, levee districts have only one: flood control. The district has the ability to tax landowners within the area, and it operates as a corporation with presidents, boards, and inspectors. Levee districts are to floods what modern school districts are to education: a function-specific unit of government that is self-governing for a specific purpose within a specified area.

Following the ideals of federalism, levee districts were scaled for the problem: they were large enough to build longer, more consistently constructed, and better financed levees than individual landowners could provide in a piecemeal way. Yet the districts were small enough to ensure local knowledge of what needed to be done. State governments had some roles to facilitate or standardize operations in their districts. For instance, state governments bought stock in levee districts to bolster their finances and also provided some engineering design standardization across districts, much like statewide education standards for schools today. But these local government units did the actual work.[3]

Levees were not limited to the Mississippi valley. Most other states adopted the levee district approach, making them a fairly common unit of local government in the Deep South. Yet devolving functions of government to the most local level can have widely variable effects, for success depends on local initiative, expertise, and diligence. Not all levee districts had the right combination. Shortly after the "forty-niners" frantically settled the Central Valley of California, the city of Sacramento formed an impromptu levee district in response to flooding in 1850. Rather than building a levee along the river, Sacramento built a levee encircling the entire town; the levee failed when the first flood hit only two years later.[4]

Competence was not the only issue; initiative was also a chronic problem for local government districts. For example, the historian John Thompson unearthed the minutes from a meeting of the Big Swan Levee District in Illinois, where the work of levee building and maintenance was often a side project for farmers. As the minutes reflect, those present were no more serious than they would be about a town hall or a homeowner association meeting:[5]

> After a long delay and much misunderstanding about when and where it [the meeting] would be held, and evening arrived and one by one the [Levee] Commissioners arrived all full of farming interests and community gossip, maybe one or more had heard some good questionable jokes.
>
> So after every subject had been exhausted and no hint or allusion to what purpose might have been in mind relative to the gathering one fellow began to reach for his hat, and another without any hat having his hand on the door knob, the Chairman suggested that we might hear the minutes of the previous meeting

if any, along with such and said bills paid or otherwise disposed of and so I begins to rummage through these notations, noting that one Commissioner had left and as I glanced up from the bill reading, I heard the lower door slam as the last commissioner had vacated the premises, and thereby assumed that the meeting was adjourned.

_____ Chairman _____ Secretary

Sign here if you ever come back.

While smaller districts in some regions of the country were flippant with their tasks, levee districts of the Lower Mississippi valley—particularly the Mississippi Levee Board of Greenville—attended to every detail of levee building and maintenance. Throughout the nineteenth and early twentieth centuries, the districts bought land for levees, built the best levees they could, and then rebuilt or repaired the levees damaged by floods. The seriousness of this task is evidenced by the steep taxes the districts set for themselves to raise funds for hiring the country's best hydraulic engineers. Thanks to the central role of levee districts in flood control, and to the burgeoning agricultural economy of the region, these districts had become nuclei of hydraulic engineering expertise by the turn of the twentieth century. The United States was completely reliant on levees for flood control, and federalism left that responsibility to levee districts.

Yet levees are not the only tool available for controlling floods. The goal of flood control is to get all the precipitated water out of a watershed in the most controlled form possible. Levees simply constrict the flow of the river to a controlled pathway and have the additional benefit of narrowing the river, which increases water velocity and makes the river deeper. Making the flow deeper and faster causes the river to further erode its own bed and so reduces the elevation of floods.

Another approach to flood control is to straighten a river by removing meanders. When rivers go along their natural crooked, tortuous paths, the water flows through turns and bends that slow down its speed dramatically. When rivers are straightened, flow speeds up, allowing more water to flow by at lower depths and thus decreasing the height of floods, all while scouring out the bed.

A third option for flood control is to use outlets or bypasses to get water out of the valley. Flood bypasses are essentially controlled, additional channels designed to take some of the river's flow and lessen the amount of water in the main channel. The Atchafalaya River is a natural outlet—a second channel—for the Mississippi to reach the Gulf of Mexico. It starts just below Baton Rouge but upstream of New Orleans. Flood control at New Orleans could use this natural outlet to decrease the amount of flow in the main channel at New Orleans.

The fourth approach is the most radical. It uses time instead of space to stretch out a flood or to control the volume of water flowing through the river at any time. This is where reservoirs play a role. Reservoirs can capture small tributary floods in the early spring and then gradually release the floodwater over a long time through the summer. If carefully planned and engineered, the release of flow from tributary reservoirs can be coordinated so that the downstream flood never reaches a critical level. Engineers often talk about reservoirs "shaving the peak off of floods."

Levees, managed by local governments, were by far the most common approach to flood control until the mid-twentieth century. Although levee districts typically requested greater state funds, and the states in turn lobbied for greater involvement from the federal government, the federal government remained distant, convinced that flooding was a local problem. During the mid to late nineteenth century, the federal government gave enormous deference to state governments. In areas where an interstate commerce interest was not clearly involved, the federal government kept the responsibility for flood control at arm's length.

Throughout the mid-nineteenth century, the Mississippi delta flooded repeatedly. The levee districts had been active in building and maintaining a growing line of levees, but their work was repeatedly destroyed by the spring floods on the river. Yet the region was a burgeoning agricultural area with growing importance in the national economy. The effect on the entire southern regional economy, along with the prevalence of New Orleans as a port of national significance, meant that the federal government felt compelled to do something about the chronic problem of flooding in the delta. Instead of doing any actual work itself on flood control, Congress chose to fund an enormous research project on flooding in the Mississippi valley as a way to nationalize the problem: the project would approach the problem in the context of the Mississippi valley, but with insights that would benefit the rest of the nation. That is, the federal government was going to fund basic scientific research.

Congress initially envisioned a single study of flooding on the Lower Mississippi that would clarify the problem and the solution. It ended up commissioning two separate studies, each proposing a course of action. The two engineers assigned to the studies were wildly different, as were the implications of their studies for the future of flood control and federalism.

Civilian engineer Charles Ellet Jr. was first out of the gate and first across the finish line. Congress had initially assumed it would fund work to be done by the Corps of Engineers, which typically was the scientific and engineering headquarters for the federal government. But the stature of civilian engineering had grown enormously by the mid-nineteenth century, as had civilian distrust of the military engineers. Civilian engineers were not just prominently involved in the levee districts of the rivers of the nation; they were also responsible for the vast canal works, locks and dams, and increasing number of railroads crisscrossing the nation. Ellet was chosen in response to the demand by civilian engineers to have one of their own do the work of evaluating the Mississippi River.

Ellet's initial career was typical for civilian engineers of that time: He cut his engineering teeth at age 17 as an assistant engineer on the Susquehanna Canal and then worked on the earliest surveys of the Chesapeake and Ohio Canal. His early training in engineering was purely hands-on amid the construction of the internal works of the nation. But then Ellet set himself apart from civilian engineers by seeking more formal education. He moved to France, the heart of civil engineering education. After a meeting with Lafayette himself, Ellet gained entrance to the *École des Ponts et Chaussées*, the world's oldest civil engineering school. In addition to his formal studies, Ellet traveled to see infrastructure that was far advanced compared to that in America of the 1830s: expansive canal networks, suspension bridges, and several massive dams and reservoirs in Switzerland.[6]

Ellet returned to the United States and developed into one of America's foremost engineers with staggering breadth of expertise. He was the chief engineer for the infant James and Kanawha Canal Company and then moved on to propose and design the first suspension bridges in America across the Schuylkill River in Philadelphia and the Ohio River at Wheeling, Virginia. Between projects (canal building and bridge building), he wrote a treatise developing a theory for establishing toll rates and freight charges to finance internal improvements first for canals and eventually for railroads. During the Civil War, Ellet shifted his attention to naval architecture and warfare and then to military strategy overall. He was one of the earliest critics of the military tactics of General McClellan, who was later relieved of command by Abraham Lincoln.

In any one of these areas, Ellet would be viewed as an accomplished scientist and engineer. Taken together, his abilities won him wide regard from civilian engineers, who hailed him as the luminary of his generation—the Steve Jobs of nineteenth-century science and engineering, capable of innovating solutions to any problem. Upon his death in 1862, many engineering magazines eulogized Ellet for his technical achievements. Almost a century after his death, the prestigious *Journal of Political Economy* remembered Ellet for his ability to think both broadly and synthetically; he was "one of those rare men of affairs who scan their activities from a general position and are capable of marshaling a seemingly incoherent mass of physical and economic data into a series of significant relations."[7]

The other engineer tasked by Congress to investigate flooding on the Mississippi River was Andrew Humphreys. If Ellet was an intellectual fox, Humphreys was the hedgehog—his life and career were bound inseparably and myopically to floods, levees, and the lower Mississippi River. Humphreys had graduated from West Point when it was still the single official nursery for the infant practice of American engineering. Humphreys graduated thirteenth in his class of thirty-one at West Point and then meandered through a series of unnoticeable posts intended for unnoticeable officers. Whereas Ellet had risen to prominence in civilian engineering circles, Humphreys stagnated in military ones.[8]

Humphreys got his break during a coastal survey, when he impressed his superior officer with his careful, meticulous work. After this relatively innocuous start, Humphreys was selected to lead the military engineers for the Mississippi River survey. He was assigned the same task as Ellet, but would represent a military engineer's view of the problem.

Not only did the two men come from different backgrounds, but their approaches were wildly different. Humphreys came at the problem militaristically and empirically; he organized three teams to study the science of hydraulics and sediment movement in the Lower Mississippi River. He collected and compiled data, working the problem like an army officer leading a platoon of data-collecting soldiers. In contrast, Ellet treated the task as if he were an academic on sabbatical; he moved to New Orleans to study the problem, bringing his wife and family instead of a team of surveyors. He relied on already available information about the Mississippi, as well as some of his own previous work on navigation from the Ohio River. Mostly, Ellet thought and wrote.[9]

In the summer of 1851, while Ellet was wrapping up his report, Humphreys had what can only be described as a mental breakdown. The physical and mental exertion of the survey had overwhelmed the army officer, and Humphreys ceased all work on the project. He first moved to Philadelphia to recuperate and then arranged to be sent to Europe to continue his recovery and examine the river works there. It was not until 1857 that Humphreys restarted the survey work of the Mississippi. By then Ellet's report had sat in congressional hands for over five years. On his return Humphreys took on an assistant, Lieutenant Henry Abbot, whom he sent back to the river to finish the remaining surveys and measurements. Finally, as the storm clouds of the Civil War gathered, Humphreys and his assistant Abbott wrote their report—a 500-page treatise on river hydraulics and the Mississippi River that would be filed to Congress just after the Civil War began.

There are many differences between the reports, but at the most fundamental level, Ellet saw the potential of the horizon while Humphreys and Abbot saw the constraints of the immediate. The Humphreys and Abbot report began, continued, and ended with sheer density, starting with the rather intimidating partial title *Report on the Physics and Hydraulics of the Mississippi River.* From there on, the military engineers laid out a masterful dissertation on the way rivers work. Their report is thick with data, tables, calculations, and footnotes. The survey had, in the words of Abbot years later, taken the "pulse of the great river," painfully measuring the intricacies of its operation upstream and downstream through all depths and all available floods. In almost every way, it represented the determinism and reductionism that would be the hallmarks of early twentieth-century engineering. Indeed, the science in the report was, and is, regarded as pathbreaking. Over the decades following, Humphreys's report established him worldwide as an authority on the science of hydraulics, and along with his conquests in the Civil War would earn him the coveted position of chief of engineers for the U.S. Army Corps of Engineers.[10]

Humphreys concluded his analysis with what became the drumbeat of his remaining career: levees were the only option. After a detailed evaluation of data and formulas, Humphreys laid out his full attack on all other opinions, condescendingly brushing aside non-levee approaches—such as upstream storage reservoirs or bypasses—

as "unrealistic." Whenever possible, Humphreys used his data to undermine any other argument, particularly those of Ellet. He would often italicize the words *direct measurement* to drive home his superior basis for authority on all matters; he had measured, others had speculated. And after hundreds of pages of his initial assault on other options for flood control, he finally arrived at his recommendation that "An organized levee system must be depended upon for protection against floods in the Mississippi valley." The reader is left with a sense of relief; the pummeling is over, and we have a solution. He then goes on to italicize his empirical, rather than intuitive, method: "It has been *demonstrated* that no advantage can be derived either from diverting tributaries or constructing reservoirs, and that the plans of [meander] cutoffs, and of new or enlarged outlets to the gulf are too costly and too dangerous to be attempted."[11]

But the essential point was not just that Humphreys had solidified the science of levees, but that he also doubled down on the existing policy of flood control: federalism. Included with his tabulations of air temperatures and elevations of historic flood stages are thirteen pages covering the history of levee district governance—a policy document within a scientific study. He pointed specifically to the levee districts of Mississippi as a "judicious State system of operations" that other states should emulate. The end result of the Humphreys-Abbott analysis was a recommendation to sustain the existing mode of federalism: the federal government should provide the science; the states should oversee systems of levee districts.

By contrast, Ellet's report could not have been more different in tone and vision: *The Mississippi and Ohio Rivers: Containing Plans for the Protection of the Delta from Inundation and Investigations of the Practicality and Cost of Improving the Navigation of the Ohio and Other Rivers by Means of Reservoirs.* Right from the start, Ellet's report was different; its title linked the Mississippi to the Ohio and identified the solution—reservoirs. Whereas Humphreys produced a treatise on hydraulics and the Mississippi River, Ellet focused on watersheds and governance and analyzed the failures of federalism.[12]

The thrust of Ellet's report was that the states of the Lower Mississippi valley, particularly Louisiana, were being negatively affected by the actions of many upstream states. Thus the entire nation had a responsibility to the downstream states. Ellet looked at the hydrology of floods and realized that planning levees around the peaks of historic floods, as Humphreys was doing, was simply naïve because past floods were poor predictors of deluges that were inevitable in the future. Ellet estimated that swamps and forested areas of river valleys upstream were themselves vast reservoirs for floods, because large floods could simply spread out onto the valley floor and be retained just as they might be in reservoirs. However, as the immense agricultural development in the Midwest continued, it would lead to the removal of forests and bottomlands upstream, which in turn would affect the Mississippi valley downstream.

For this reason, Ellet took the federal government itself to task. Until 1849 undeveloped swamplands, lying mostly along rivers and floodplains, were in the public domain. They were owned by the federal government, similar to U.S. Forest Service lands of the twenty-first century. To stimulate land development and settling of the interior, Congress passed the Swamp Land Acts of 1849, 1850, and 1860, which transferred ownership of nearly 65 million acres of swamplands in fifteen states to the governments of those states. This transfer of ownership had a caveat: funds from the sale of these lands by the states had to be used to build the levees and drainage required for land development. The federal government was not only encouraging settlers to move into flood-prone lands but also requiring deforestation of upstream swamps and floodplains of the upper Midwest and thus increasing the flood load on the downstream Lower Mississippi valley. Ellet saw the federal government as part of the problem and said that it had to be part of the solution as well.[13]

Because the downstream floods were so crippling, and because they would only get worse, Ellet's solution demanded that anything and everything should be tried. He argued for opening river outlets on the Lower Mississippi River to Lake Borgne and enlarging outlets to Bayou Plaquemine and the Atchafalaya River. Ever mindful of the effect of upstream actions on downstream communities, he was against straightening the river not because he thought the result would be ineffective, but because he worried that doing so would exacerbate flooding downstream. Most importantly, Ellet advocated for reservoirs. He suggested establishing a system of upstream reservoirs in the tributaries of the Ohio and Mississippi. The reservoirs could store floods in the spring and then release the flows gradually during the summer months to minimize floods as well as supplement low flows for navigation. The reservoirs Ellet envisioned were nontrivial, estimated to be staggering in size and costs. But Ellet believed that reservoirs were essential. Levees had a role, but they should be used cautiously and as a last resort. Relying solely on levees, Ellet thought, gave "a delusive hope, and most dangerous to indulge because it encourages a false security."[14]

In 1862, only months after Humphreys submitted his report—but years since Ellet had submitted his—Ellet was killed on the Mississippi River at the First Battle of Memphis. Humphreys, in contrast, rose to fame during the Civil War. He went on to serve as chief of engineers for the entire Corps, where he would lead the implementation of his own recommendations. The centerpiece of his recommendations, and therefore the centerpiece of Corps flood control policy, was a strict "levees only" doctrine for flood control. The Corps moved gradually into the realm of flood control, but only on its own terms: levees.

Over the ensuing decades, levees grew in size and number, expanding well beyond the Lower Mississippi valley. By the start of World War I, fifty-two levee districts had been created on the Upper Mississippi, protecting several hundred thousand floodplain acres, and over 330 miles of levee were built along the Illinois River. Several hundred more miles of levees were built to protect the agricultural lands of the Sacramento Valley in California.[15]

Ellet had lost the flood control battle to Humphreys for ideological reasons as much as scientific ones. Humphreys's support for state-driven, levee-centric flood control was a logical extension of late-nineteenth-century federalism; Ellet's proposed solutions—requiring systematic planning on a national scale and substantial financing by the federal government—would have necessitated reorganization of federalism itself. If there was going to be any approach to flood control other than building and maintaining levees under state government oversight, it would require some kind of event or events that undermined the nation's perception of levees and federalism. A series of mega-floods provided one such event, and the Great Depression provided the other.

CHAPTER 4

Flood Control

W hen the spring rains began in 1927, farmers along the Mississippi were in their typically precarious position of estimating the odds of losing that year's crop. On the one hand, flood control had improved: there were over 250 million cubic yards of earth forming the levees of the Lower Mississippi—enough to bury 240 square miles under a foot of earth. Levee district engineers were widely recognized scientists who traveled internationally and wrote articles for prestigious scientific journals. Levees were also more effective: from the close of the Civil War to 1927, the chances that a flood would inundate farms along the river had been cut roughly in half.[1]

Yet booms in population, industry, and agriculture had made regions along major rivers all the more vulnerable to the impacts of levee failure, and the levees—along with floodwalls in urban areas—were far from perfect. The 1897 Mississippi River flood inundated 87 percent of the area protected by the Mississippi Levee Board (2,270 out of 2,593 square miles). And after considerable rebuilding, the levees failed again in 1912, inundating over 52 percent of the district's protected area. Urban areas bore no less impact. The flooding of Pittsburgh's central business district in 1907 put 100,000 workers out of work amid growing labor tensions. The subsequent 1913 flood on the Ohio River was 15 feet above any previously recorded heights, caused over $300 million in damages, and killed more than seven hundred people—more than had died in the Great Chicago Fire of 1871. The growing population in California was also experiencing devastating floods as the Sacramento River continued to surprise everyone who worked on it. Local engineers at the turn of the twentieth century estimated that the peak flooding the Sacramento could ever experience would be 300,000 cubic feet per second (cfs); in both 1907 and 1909, the Sacramento neared 600,000 cfs.[2]

But no flood had greater effect on the psyche of a region, and the nation, than the 1927 Mississippi River flood. Almost any description or statistic of the 1927 flood seems over the top, but no single account fully conveys the severity and scope of the disaster. One hundred breaks in the mainline Mississippi levees inundated over 26,000 square miles across seven states—a region larger than the total area of West Virginia. Over 700,000 people lost their homes and as many as 300,000 were rescued from houses, rooftops, levee crowns, and even trees. Tens of thousands of livestock were killed, and millions of acres of the nation's most productive farmlands were destroyed.[3]

Beyond the direct impacts, farmers in the Midwest were dependent on the Mississippi to export their products through New Orleans. With the river and region crippled by the flood, all the agricultural exports were bottlenecked in the Upper Midwest. For weeks, trains were unable to cross the Mississippi south of St. Louis because more than three thousand miles of track were underwater. Total losses were diffused across the nation and hard to estimate, but they reached as much as $1 billion at a time when the federal budget was typically less than $3 billion. As agricultural commodities spoiled and wasted in Missouri and Kansas because of flood waters in Louisiana and Mississippi, flood control began to appear more and more a federal rather than local issue.[4]

The flooding didn't stop in 1927. Starting in 1928 in Florida and continuing for the next decade throughout the United States, floods pummeled just about every region of the country. In the late summer of 1928, floods in central and south Florida killed 2,500 people, mostly poor blacks in the Everglades.[5] Then, beginning in January of 1935, fatal floods hit the state of Washington, then the James River in Virginia and the Kanawha River in West Virginia, killing dozens of people and causing millions in damage. Later that spring, floods in Kansas and Texas in turn drew the attention to the middle of the country; then floods wracked New York and parts of the Midwest along the Ohio River. All told, more than two hundred people died in the floods of 1935 and damages topped $130 million. Less than a year later, floods began wreaking havoc yet again throughout the Northeastern United States. Rivers there reached levels previously thought unimaginable: the Connecticut River crested almost nine feet higher than any flood recorded since Europeans had settled there in 1639. Throughout Pennsylvania, 82,000 buildings were destroyed, and Pittsburgh sat under 16 feet of water on Saint Patrick's Day with 47 dead and 67,000 homeless; Pennsylvania would later estimate total damages at a shade over $210 million. Mother Nature seemed to drive home the point by flooding Washington, D.C., for good measure: the Potomac River rose almost 20 feet above flood level, submerging the capital's riverfront parks and requiring thousands of sandbag levees to protect the Lincoln and Washington Monuments.[6]

These floods ended the era of state-centered federalism for flood control. With the Flood Control Act of 1928, the federal government got into the business of flood control. The 1928 act was staggering: It authorized spending $325 million ($4.5 billion in 2010 dollars) on flood control on the Mississippi and Sacramento Rivers, making it the largest public works project ever considered—exceeding even the $310 million authorized for building the Panama Canal. This act was focused on two specific rivers that had some connection with interstate commerce, thus allowing Congress to keep flood control under the umbrella of navigation. But when Congress passed the subsequent Flood Control Act of 1936, everything changed: not only did Congress authorize $310 million to fund the work, but Section 1 of the act opened with a "Declaration of Policy" stating that "it is the sense of Congress that flood control is a proper activity of the Federal Government." That is, the federal government was now fully in the business of flood control. And the Corps was the arm of the federal government doing the work, becoming the nationwide planning and construction agency for flood-prone rivers.[7]

Whereas the 1928 Flood Control Act had focused on the Mississippi River as well as some work on the Sacramento, the 1936 act was a truly national bill in that it sought to address flooding across the United States. More importantly, it also wholeheartedly adopted reservoirs, floodways, and cutoffs as valuable methods of flood control in addition to levees. Indeed, almost two-thirds of the funds the bill provided were earmarked for reservoir construction in New England, New York, California, and tributaries of the Mississippi River including the Ohio River (and its tributaries) and the Arkansas River. Every major river in every region was going to have a reservoir for flood control.

That so many floods hit during the Great Depression was an essential element of changing the federalist calculus of flood control. The 1930s brought two essential changes in mind-set. First, the enormous costs for the infrastructure were to be federalized in the name of absorbing labor, and flood control projects were ideal due to the sheer scale of the projects and the number of jobs they absorbed. The building of reservoirs across expansive watersheds meant that many different regions of the nation got projects, and workers across the United States got jobs. Because much of the work was simple manual labor, poor workers could be drawn from the rural, impoverished regions that needed jobs. Second, the 1930s were the apex of the Progressive Era, when leaders were enamored with systems planning, optimization, and engineering. Charles Ellet's vision of the flood problem as a watershed, system-wide problem dovetailed with the Progressive ideology of the time: the system had to be optimized, which meant that a central agency—the Corps of Engineers—had to be in charge of planning, engineering, and coordinating.

Large river basins quickly became peppered with Corps-designed reservoirs and floodways or bypasses. Levees and floodwalls would remain critical parts of the system—and be expanded—but headwater reservoirs would store spring floods and then release the flows gradually during the summer, thus subsidizing low flows for navigation, irrigation, or hydropower. And when needed, floodways would be activated to divert some flow out of the main river. River systems became highly engineered, optimized hydraulic machines. Early twentieth-century floods gave the motivation; the progressives gave the ideology, and the New Deal provided the resources.

Seventy-five years after the 1936 act, in 2011, Kent Parrish at the Corps and Peter Nimrod at the Mississippi Levee Board managed their parts of this system while faced with a 1927-scale flood. Each organization had its own roles and responsibilities that drew on the advantages of each level of government. The flood of 2011 demonstrated how the system worked when it worked very, very well.

In the decades since the federal government first moved into flood control, it has implemented Ellet's vision across the Lower Mississippi River. The system that emerged converted the Mississippi's natural flows and variability into a conceptualized pipeline with valves and tanks—bypasses and reservoirs—to divert and store the worst-case scenario of the projected flood. Under that scenario, the various tributaries would be delivering their maximum or near maximum flows at the same time as other tributaries. To manage the flows, reservoirs would store water while bypasses diverted water in precise flows at precise locations.

This is the system of river management that now exists, and it is the one Kent Parrish thinks about when he joins conference calls with Corps personnel up and down the Mississippi and Ohio Rivers, as well as with hydrologists from the National Weather Service. In the late nineteenth and early twentieth century, levee engineers had to guess how high the river would get each year based on haphazard telegrams from upstream sources and their own gut instincts acquired through years of living along the river. Federalizing the river systems, as is the case in the twenty-first century, dramatically improved the availability of information. Through their complex system of weather stations, flow gauges, and computer models, the combination of federal agencies—the U.S. Geological Survey, the National Weather Service, and the Corps of Engineers—now forecast with remarkable accuracy how high the river will rise at every city along the river.

In spring of 2011, rain had been coming down in the Ohio River basin for weeks and weeks, as well as in the Upper Mississippi. Recalling the events two years later, Parrish remembered the atmosphere in the conference room when he and his engineers sat down for the interagency call on Good Friday in 2011. That was the call when the National Weather Service dropped the bomb of their forecasted river stages (water-level heights): 59 feet at Greenville and 53.5 feet on May 13 at Vicksburg. "When they said that number," Parrish remembered, "My jaw dropped." Someone else in the room simply let out a long whistle. A flood stage measuring in the mid-50s put this potential flood beyond the floods of 1873 or 1973 and into the orbit of 1927. It turns out that they underestimated; the Mississippi River eventually crested at Vicksburg on May 19 at just over 57 feet.

But more important than accuracy of the information itself, the federal government now had infrastructure that was all working as a system in a kind of managed, organized way. Or, as Parrish put it, "You gotta coordinate this stuff." Given predictions of the upstream flood wave as it made its way downstream, Corps reservoirs on major tributaries such as the Cumberland and Tennessee Rivers were being cleared of water ahead of time. The idea was to reduce their discharge for later when the Ohio was predicted to crest. Thus, as Ellet had suggested over a century earlier, the coordinating reservoirs successfully reduced the total load on the system.

Nevertheless, the water level was still rising in the Lower Mississippi. Once the forecasting and reservoir operations and bypasses had been used, flood control came down to the local entities that bear the brunt of fighting floods. And this is where levee districts come in.

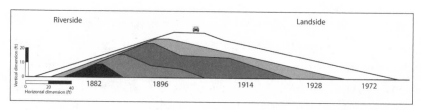

The Mississippi River levees were built up dramatically between 1882 and 1972.

The work of the local flood districts always seems to involve sand-bags. Before levees on the Mississippi River were extended to their current height, engineers lived in chronic fear that the river would overtop the levee, thus destabilizing it and causing it to fail. To avoid overtopping, sandbags were used to add a few more feet to the height of levees. However, now that the levees are impressively high, the chances of overtopping are smaller, and sand boils have become the greatest threat to levees. When rivers are at flood stage, the river itself is dozens of feet higher than the land on the other side of the levee. This creates enormous hydraulic pressure on the levee itself: the pressure exerted by the Mississippi River on the levee at flood stage is about the same as that exerted by a school bus with all its weight pressing on a single square yard of area. Because levees have been built and rebuilt to resist this pressure, water instead finds that going through the soil beneath the levee is the path of least resistance. If the water flows too quickly under the levee, it can pull and flush soil—often sand—along with it. This creates a sand boil on the landward side of the levee because sub-levee water flow is effectively eroding sand beneath the levee. If the sand boil continues, along with the soil erosion underneath, the levee will collapse under its own weight.[8]

The way to fight sand boils is to increase water pressure on the landward side to counteract the pressure of the water flowing underneath. This is done by encircling the sand boil with sandbags. As the water rises within the ring of sandbags, it begins to look like a swimming pool; once it gets deep enough, the pressure equalizes and slows the flow of water beneath the levee.

Sandbagging sand boils is the work of local levee districts. It is the last line of defense—trench warfare—against a flood. During the 2011 flood, levee districts in Louisiana alone placed more than a million sandbags around boils. On the Mississippi side, Peter Nimrod and others from the Mississippi Levee Board continuously checked every foot of the 212 miles of their levees in search of boils or other weaknesses. When sand boils emerge, the levee district has to find the labor to take care of them immediately. Mobilizing sandbag crews, often in the middle of the night, is a large part of Nimrod's job at the levee district.

Most people, if they ever handle a sandbag, do it when the sand is dry. But flood fighting is wet, noted Nimrod: "Sandbaggin' in a

A sand boil near Greenville during the Mississippi River flood of 1927.

flood is moving sandbags wet; imagine how heavy they are; about 65 to 70 pounds." Each time a sand boil appears, the levee crews—often including prisoners who work in exchange for time off their sentences—dash to it and ring it with sandbags. They often do this backbreaking work the entire night while the river looms four stories above them on the other side of the levee. Standing at the remnants of a sand boil that they fought during the 2011 flood, Bobby Thompson from the levee board described how he felt during the many days and nights of the flood fight: "I wasn't scared. I was just tired. I never been in combat, but I don't know how real fightin' can be more tirin' than flood fightin'."

This mundane, on-the-ground, backbreaking work is part of twenty-first-century flood control. It involves a staggering array of governments and responsibilities: NASA satellites report precipitation, and the National Weather Service forecasts flood stages; Corps of Engineers generals zip around by helicopter to optimize reservoir levels and operate bypasses and floodways; state governments organize evacuations of the people living in areas that might be flooded; and local levee districts up and down the river do the nitty-gritty work of ensuring the structural integrity of the levees. This system is all the result of a layered approach to federalism that in many ways has worked. On May 8, 2011, the Mississippi River gauge at Greenville, Mississippi, topped 60 feet for the first time since 1927; yet the levees held while the upstream reservoirs peaked at their maximum storage capacity. Indeed, the discharge of the Mississippi River from Cairo, Illinois, to Baton Rouge, Louisiana, was higher in 2011 than it had been in 1927, yet flooding caused minimal damage.[9]

Despite the levee managers' enormous successes in 2011, when I visited two years later, their attention was already focused on their next potential project—a gargantuan pump. This pump, known as the Yazoo Backwater Project, was envisioned to send enormous amounts of water out of the Yazoo River area of the delta, over a levee, and into the Mississippi River. While much of the project had been constructed, including the levees, the final piece was the not-yet-constructed pumps. Both Kent Parrish and Peter Nimrod spoke of the pump topic repeatedly as we toured levees of the region. The men were frustrated by not being able to get money appropriated from the federal government to build this project that they felt was so plainly needed. As they pointed out, Congress had authorized the project in the mid-twentieth century; but lawmakers hadn't yet appropriated funds to pay for the work. Parrish and Nimrod were frustrated that the federal government would not fund another part of this clearly successful infrastructure project that had been part of the broader projects protecting the Mississippi delta. Perhaps more cynically, they were trying to bring home some political pork as part of protecting their region from floods. And they were not alone.

Although the need to control floods had brought the federal government into the business of dams and levees, the need to absorb labor during the Depression had justified the federal government to spend freely on flood control infrastructure. When states and levee districts were economically crippled, the federal government stepped in and paid for these projects, largely to provide jobs but also to get large projects built. But like many other aspects of river infrastructure during this era, projects didn't end once the Depression did. The pendulum of federalism had swung dramatically toward the federal government being the locus of power, and there was little desire to go back to the way things were.

Flood control infrastructure projects became a favorite flavor of political pork for the next half century. Congressional leaders found they could deliver local benefits at federal expense, and the Corps provided an unending list of projects for Congress to fund. The 1936 Flood Control Act kick-started the flurry of activity and led to the Corps constructing hundreds of flood control reservoirs along with tens of thousands of miles of levees. Eight years later the 1944 Flood Control Act allowed for the construction of mammoth dams, along with their accompanying reservoirs, on the Missouri River: Fort Peck,

Garrison, Oahe, Big Bend, Fort Randall, and Gavins Point together inundated over 750 miles of river and stored almost 74 million acre-feet of water in a string of human-made lakes that is longer than Lake Michigan and Lake Superior together. Subsequent flood control acts followed rapidly and approved projects in every corner of the United States that had not gotten a project from the 1936 act. From the founding of the United States until 1928, local governments had borne the bulk of flood control costs; but from 1928 to the close of the twentieth century, the federal government took up spending. This was the era of minimum federal restraint for flood control.[10]

And then, just as dramatically as it had begun, the pendulum of federalism swung back. For flood control infrastructure, the pendulum started its dramatic reversal on November 17, 1986, at the pinnacle of the Reagan Revolution and the president's efforts to decrease the federal government's role in almost every aspect of society. For flood control, this meant putting the impetus back on local and state governments to spend their own money on projects. The Water Resources Development Act of 1986, passed in November, required local sponsors—local districts or state governments—to pay half the costs of projects. Proposed projects, even those already authorized by Congress, came to a screeching halt. Local levee districts and even state governments found they didn't desire many flood control projects nearly as much when they had to pay 50 percent of the often exorbitant costs. Indeed, this was the case with the Yazoo Backwater Project in Mississippi; the project had been authorized in 1941, and some components had subsequently been constructed. But after the 1986 act, construction of the final pump station was postponed indefinitely because the local district could not afford its share of the costs. Projects across the United States faced similar fates. From 1986 on, the era of massive flood control infrastructure was largely over; federal restraint on spending and shared local responsibility proved enough to stanch the hunger for constant construction of flood control mega-infrastructure.[11]

Infrastructure was the first flavor of political pork for flood control. But with federal coffers shut for infrastructure, pork seekers began looking for other sources of federal funds, and they found a vast supply in the form of disaster relief. Just as building flood control infrastructure is not necessarily greedy or poor governance, providing funds for disaster relief can create similarly perverse incentives when not disciplined by federalism. And just as Charles Ellet anticipated the problems of the flood control system in a government report, the emerging problems with disaster relief would be exposed in another scientific report on flooding—this time written by a brigadier general from the Corps of Engineers.

Gerry Galloway is blue-blood Corps. His father was a West Point grad who then became the commandant of engineers at West Point and retired as a major general after commanding the vast Army base at Fort Belvoir outside of Washington, D.C. Following in his father's footsteps, Gerry graduated from West Point and joined the Corps. He rose quickly through its ranks while picking up a master's degree from Princeton in engineering, another from Penn State in public administration, and a Ph.D. in geography from Chapel Hill. His early and middle career in the Corps included one of the most prestigious appointments; he became commander for the Mississippi Valley District, managing all federally constructed mainstem levees through the delta.

But then Galloway pivoted and took a different trajectory. Rather than continuing in the Corps hierarchy and potentially becoming chief of engineers like Andrew Humphreys a century earlier, Galloway left the Corps—in a way—by joining the faculty at West Point. His career there was no less illustrious; he quickly became the dean and chief academic officer at West Point and was promoted to brigadier general. Through his role at the Corps as a practicing engineer and then his role in training future generations of engineers at West Point, Galloway became known among river managers in the United States as a flood control engineer; when his name is brought up in

Iowa or Mississippi, levee district managers—including Nimrod in Greenville—inevitably know who he is and refer to him simply as "the General." But Galloway was also outside the normal Corps in the early 1990s; he had the bona fides and blue blood of a career Corps man, but he had also spent a portion of his time at the margins of the typical Corps—part academic, part Army engineer.

In 1993, perhaps because of this insider/outsider status, President Clinton tapped Galloway to analyze the aftermath of a massive flood in the Midwest. That year had been the wettest summer in 100 years for much of the Midwest. More than 1,800 miles of rivers and tributaries were inundated by record high flows, including over 500 miles of the upper Mississippi, 400 miles of the Missouri, and almost 200 miles of the Iowa and Kansas. Every river and every tributary was swollen with water and stayed that way, so that flooding became almost normal. The Missouri River at St. Louis first went above flood stage in early April and would not fully go back down until mid-October. Over 26 million sandbags were used in fighting floods in 1993, but to little avail: as rivers went over their banks, they eroded the roads, along with millions of tons of topsoil as the flooding broke levees and raged across the steep, narrow Missouri River floodplain. In some places the river scoured massive divots in the floodplain, creating a landscape that resembled an area bombed during a war. In other places the flooding river blanketed farms with several feet of sand, effectively sterilizing the fields. The 1993 flood far exceeded the 1927 Mississippi flood in spatial extent and economic impacts: it inundated over 20 million acres as compared to the almost 13 million acres flooded in 1927.[12]

At the time of the flood, Gerry Galloway was at West Point. Twenty-three years after the flood, Galloway—now retired from the Corps and West Point but still working more than full-time for the University of Maryland—recalls the 1993 flood as "the first CNN flood." It was also the first natural disaster of the Clinton administration, which tapped Galloway to wrestle with floods in the United States. The general went on leave from West Point and was given a team of experts to work with him in office space in a brownstone right on Lafayette Square, facing the White House. But Galloway's task was not to oversee the rebuilding of infrastructure, the operational aftermath, or disaster relief. Rather, he was brought in to untangle the causes and consequences of the flooding and evaluate the government—federal, state, and local—responses to floods. Like Humphreys and Ellet over a century earlier, Galloway was asked to take a step back and rethink how floods were being dealt with.

When recalling where he started to tackle the problem, Galloway went to the immediate aftermath of the 1936 Flood Control Act: "After thirty-six, nobody asked whether we should continue developing and building in the wrong areas." This was the central sentiment in the Galloway team's approach to the 1993 flood, and for the remainder of Galloway's career: Before talking about flood control, should we be living and working in floodplains in the first place?

In the late twentieth century, flood control infrastructure had created an enormous sense of hydrologic security. But it also unintentionally created a perverse incentive that drew people into areas previously considered too high risk for developing. Levees initially constructed to protect farms and a few rural houses were now protecting neighborhoods, industrial parks, and portions of cities. Flood control infrastructure began to be economically justified based not on protecting what actually existed, but on what could exist if floods were eliminated. When evaluating a potential flood control project, a cost-benefit analysis may show the project was not justified if used only to protect farms. But if flood control led to new industry and housing developments replacing the farms, then the cost-benefit ratio for justifying projects could be flipped in a favorable direction to support the building of new or expanded infrastructure.

This new logic had developed after the 1936 act and was prevalent throughout much of Galloway's career in the Corps. It created a tautologous rationale that led to flood control infrastructure and floodplain development being codependent—working together to encourage and even require development in extremely hazardous areas. Not only did proposed flood control projects increase the likelihood of such development, they depended on it. The combination of levee districts, state governments, and federal engineers building mammoth reservoirs and precisely engineered levees created the appearance and assumption that flooding had been eliminated. Society came to believe that the floodplains could be made flood-proof—and so, even though intense investment in flood control reduced the number of floods nationwide, economic impact of floods increased annually as more and more people lived on floodplains. Because of this shift, more infrastructure went up along floodplains, thus drawing more people to live on floodplains. And when events like the 1993 flood took place, they were disasters of staggering scale despite—or perhaps because of—the growth of flood control infrastructure.[13]

The federalism question of the early and mid-twentieth century had been "What should be the role of the federal government in preventing floods?" The federalism question of the late twentieth century became "What should be the role of government during and after floods?" The United States lacked an overarching policy or specific program for responding to natural disasters, but it did provide some financial relief following specific, significant floods like the 1927 flood. In 1950, however, Congress authorized a relatively minor bill to fund the repair of roads and bridges that had flooded along the Red River in Minnesota. This small act was a big deal: its funds were solely for reconstructive work, and its funds were solely for local work. From that precedent the federal role in disaster relief ballooned like the federal role in infrastructure had before it. By 1980, disaster response and relief had been reorganized into a new administrative home—the Federal Emergency Management Agency (FEMA). Through FEMA and growing spending and responsibilities, the federal government took on the role of providing temporary housing, grants for repairing damaged state property, unemployment compensation, health services, and payments to communities to offset lost tax revenue.[14]

In addition to funding these purely local effects of flooding, the federal government also developed questionable practices for evaluating disasters as being eligible or ineligible for relief funding. A governor from an affected state would request a disaster declaration from the president; if the request was granted, the state gained access to federal funding. These disaster declaration requests were routed through FEMA, which was outside normal cabinet-level oversight, so that their review was responsive only to the White House. Along with this lack of systematic agency oversight, the Stafford Act in 1988 explicitly prohibited the use of "arithmetic formulae" (such as benefit-cost analysis) as a basis for disaster declarations. Disaster relief became codified as a solely political decision, outside traditional economic evaluation. Because disaster relief allows federal politicians to helicopter into stricken areas with the promise of federal funds close behind, for the White House a disaster request from a governor was an attractive request to approve: the Clinton administration declared 709 disasters, or about 1.7 disasters per week; the Bush Jr. administration declared 1,037 disasters, or 2.5 per week; and the Obama administration approved 938, or 2.25 per week.[15]

Federalizing disaster relief in turn reduced the burden carried by local and state governments: federal disaster relief required only a 25 percent *non*federal cost share, thus ensuring that the federal government would bear most of the relief costs. Yet this effort at local and state responsibility was consistently undermined: the nonfederal share was reduced or waived altogether following many severe disasters, and many local and state governments avoided most disaster relief costs altogether. Few local governments were calling for disaster funding reform, and essentially no sustained calls for fiscal discipline came from federal leadership because no senator or congressional representative was certain that his or her constituents would escape flood damage during the next season. The sprawling federal flood control infrastructure program of the mid-twentieth century had been replaced by an equally massive federal disaster relief and recovery assistance program of the late twentieth century.[16]

Even the few efforts by the federal government to correct this series of systemic mistakes backfired. When floodplain property owners were left without insurance options because the private insurance industry had little appetite for the enormous risk created by river floods, the federal government stepped in with a new low-cost insurance program: the National Flood Insurance Program (NFIP). But NFIP was the only insurance option for properties in flood-prone areas, meaning that the federal government was insuring only high-risk properties. With such high-risk exposure, NFIP payouts were higher than the revenues taken in by the program, so the program had to borrow chronically from the federal treasury. The unintended effect of NFIP was to federally subsidize the costs of living on floodplains.[17]

This entire flood control and disaster relief system drew the scrutiny of Gerry Galloway as he and his team sequestered themselves to analyze the causes and consequences of the 1993 Missouri River disaster. The existing infrastructure—all the massive dams upstream and levees downstream—had reduced much of the damage that otherwise would have occurred. But Galloway's analysis started to reveal that insurance and disaster relief programs were especially problematic. Only about one in five properties damaged by the 1993 flood were actually insured; and while the NFIP paid out over $293 million, this amounted to less than 7 percent of the $4.2 billion of direct federal disaster relief costs. Even the cost-share efforts that had been introduced to increase local responsibility were undermined. In response to the flood, a Housing and Urban Development (HUD) program provided $450 million in grants for flooded communities to use as they saw fit. Some communities turned around and used this HUD money as their "nonfederal share" of flood disaster recovery, thus routing federal funding from one agency to provide the nonfederal funding of another agency.[18]

Galloway and his team chronicled these types of examples in high detail across the flood-affected region. Then they summarized and analyzed them in a report that reads more like a treatise on flood policy than a government report. This report to the White House— eventually known as "The Galloway Report"—was one of the first systematic analyses and criticisms of flood control and floodplain management in the United States. It was a full frontal analysis, and maybe even assault, on assumptions and practices that had accrued over the decades since Humphrey's report became the bible of the Corps. Although the Galloway Report blamed the flood damages squarely on Mother Nature, calling the Midwest flood a "significant hydrometeorological event," it also blamed the impacts of the flood on the demise of federalism: "Through provision of disaster assistance, the government may in fact be reducing incentives for local governments and individuals to be more prudent in their actions." The thrust of the report was for the federal government to exhibit restraint so that local governments would exhibit initiative.[19]

The report was also one of the first to describe federal flood disaster relief as a moral hazard. When one person decides how much risk to take on while another person bears the costs if things go badly, the result is a moral hazard. Insurance companies have long worried about moral hazards created by overgenerous reimbursements: if their insurance policies were too generous, then their clients might behave in systemically risky ways. Similarly, in finance, the federal bailout of banks that were deemed too big to fail during the 2008 Great Economic Recession created a moral hazard: banks could grant risky loans and enjoy high returns, yet the U.S. Treasury bore the costs of those risks. This system created a moral hazard of giving banks an incentive to engage in risky behavior.

Galloway suggested that the existing federal disaster relief program created a nationwide moral hazard for floodplains. By providing artificially low-cost insurance along with disaster relief to entire communities, the federal government was encouraging risky decisions by individuals and communities on floodplains. Indeed, the most damning components of the Galloway Report focused on federal insurance programs as the root cause of the moral hazard. While only 20 percent of the flooded properties had flood insurance, over a third of those policies were purchased within a few days of the property being flooded—because the federal insurance program had only a five-day waiting period for claims. Those living along a river simply waited and watched the weather and flood reports over the winter and spring to see if they might be flooded. Then, within a week of the flood peak reaching their part of the river, they purchased insurance. They later received full payments from the federal insurance coffers.[20]

In the same way that Ellet took the federal government to task for its role in mid-nineteenth-century flooding of the Mississippi delta, Galloway took the federal government to task for exhibiting little restraint following almost any natural disaster. During the three natural disasters that occurred in 1993—the Midwest flood, Hurricane Andrew, and the Northridge, California, earthquake— nonfederal cost-share requirements decreased while disaster costs increased. Consequently, insurance programs went further and further into debt. This weaning of local responsibility perpetuated the view that the federal government would always be a benevolent big brother with ever-deeper pockets—a financial safety net. The end result was to encourage further development on dangerous floodplains.

Galloway was not a lone voice. An increasing number of scientists, engineers, and planners had likewise begun arguing for a first step of federal restraint. Perhaps more surprising, some state and local communities and governments—those benefiting directly from federal disaster relief—began wanting federal involvement reduced rather than enlarged after experiencing the moral hazard effect. Following a series of floods along North Carolina's rivers in the 1990s, an editorial in the *Raleigh News & Observer* chastised the federal government for its role in creating the moral hazard at taxpayer expense: "The allocation of hundreds of millions in taxpayer dollars has led the federal government to undermine what state officials have been trying to do for decades—discourage development in floodplains."[21]

Over the next decade, the Galloway Report became famous among engineers and planners as the kind of thoughtful document that sheds light on systemic problems and systematic solutions. Yet, in the Corps of Engineers and even within the federal government,

Galloway did not benefit directly from his analysis and critique. A key implication of his argument was that floodplains might be better suited as environmental corridors—natural ecosystems of swamps and wetlands—rather than as shopping malls and suburban sprawl. Environmental groups praised Galloway's report and its insights, but land developers and levee districts began to question his dedication to flood control. In 1995, a year after the Galloway Report was published, Galloway's name was floated for the position of Assistant Secretary of the Army for Civil Works—the chief political appointee over the Corps, and an appropriate placement for someone widely regarded as a luminary of river policy. Yet traditional flood control interests were nervous about Galloway's acquired green label, fearing that he would undermine the status quo on which they depended. When Galloway was passed over for the appointment, the *Washington Post* noted the irony that the hint of an environmental label was enough to sink the nomination of an active duty brigadier general at West Point.[22]

After being passed over for Assistant Secretary, Galloway moved on. He became dean at the National Defense University and then in 1998 formally retired from public service. For the next two decades, General Gerry Galloway spoke on the radio or television to give his opinions about why the costs of floods were increasing rather than decreasing. He called for federalism to be applied more stringently— reducing the federal role and increasing local responsibility for flood control and disaster response. He occasionally penned editorials to revisit the recommendations of his earlier report. And he pointed to places that were particularly vulnerable to catastrophic flooding— places like New Orleans.

Of all the disasters, Hurricane Katrina has the dubious distinction of revealing the chaos of federalism, flood control, and moral hazards on live national television. If the 1993 Midwestern flood was the first to be called a CNN flood—for media coverage of the aftermath of a flood—then Katrina was the first flood to be covered as it took place. But in many other ways, the flooding disaster of Katrina is simply a continuation of historic trends, although focused in a small place for a short, intense period of time.

Like other cities prone to flooding, New Orleans has a complicated hydrology that convolves multiple sources of water with flatness. The city proper is wedged between the Mississippi River to its south, Lake Pontchartrain to its north, and Lake Borgne to its east (the lakes actually are tidal inlets to the Gulf of Mexico). Floods thus come at the city from all sides: from the river upstream, as in the case of the 1927 flood, or from coastal surges up into the lakes, as happens during hurricanes.

The original city was built on a topographic anomaly—a small natural levee on the north bank of a sweeping meander of the Mississippi River. This slight rise has stayed surprisingly dry through floods of the past, largely because the river had so much area to

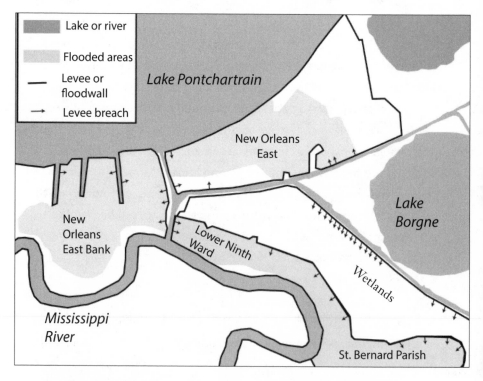

Levee breaks and flooding in New Orleans during Hurricane Katrina.

flood elsewhere: the river could send much of its flow west through the Atchafalaya basin over two hundred miles to the northwest, release more of the flow through Lake Pontchartrain just north of the city, and pass a smaller amount downriver past the small hill of downtown New Orleans. These natural outlets to the Atchafalaya and Lake Pontchartrain, however, were closed off during the Corps' levees-only era in the late nineteenth and early twentieth centuries. Then the Mississippi was allowed to flow only through its channel past New Orleans. The city thus became fully dependent on the levees for passing floods. The other significant geographic turn was suburban sprawl. Starting after World War I, the population of New Orleans began spreading outward from downtown, off the high terrain of the natural levee toward lower-lying areas near the lakes, particularly Lake Pontchartrain to the north.

As the city and region have grown, two trends have been at work to ensconce the city in its own flood control infrastructure. First, a combination of factors—some natural and some human-induced— are causing subsidence; the city is literally sinking deeper into the swamp. The original downtown remains on a slight rise above sea level, as in the early eighteenth century. But an increasingly large portion of the expanding city outside downtown continues to sink below sea level, particularly the area known as the Lower Ninth Ward to the east of downtown and those areas to the north along Lake Pontchartrain. With large swaths of the city below sea level, rainwater and groundwater seep relentlessly into the marshy soil of the city—hence the need for aboveground tombs rather than belowground graves in the city's cemeteries.

In addition to being at sea level and the delta of a massive river, New Orleans is located in one of the wettest places on Earth, and direct rainfall often causes flooding. Average annual rainfall exceeds 50 inches per year, and extreme events can be downright shocking: in 1927, before the Mississippi River flood crest reached New Orleans, the city received 18 inches of rainfall in 48 hours. To stay afloat, the entire city has to be dried through a series of drains and canals that collect rainwater or seeping groundwater. Enormous hydraulic pumps work to pump this ever-seeping water up over levees and into canals that run from the edge of downtown and drain north to Lake Pontchartrain. Without those pumps, local groundwater and rainwater have nowhere to go. They would accumulate and effectively turn the below-sea-level city into a massive puddle.

At the turn of the twentieth century, a substantial wetland fringe existed between the city and the shores of Lake Pontchartrain to the north, but the city began creeping into this bottomland. A local levee district built a seawall offshore in the lake and then created dry land behind the wall by pumping sand from the lake bed up and over the wall. As the wetlands along the canals drained, more dry land was available to develop the growing city. The city thus expanded northward toward the lake into former wetlands. Canals draining water from downtown toward the lake were lined with single houses; rather than passing from downtown through wetlands toward the lake, the canals passed from downtown through newly settled suburbs where land was chronically subsiding. Low ridges of dirt initially built to keep drainage flow in the canals instead became critical flood control infrastructure for these new portions of the city as it sprawled into the former wetlands and lake bed. As more homes were built and demand for land increased, it was more and more difficult to expand the footprint for levees; building levees higher meant expanding their base foundation, which consumed land that could otherwise be used for development. Instead, sheet pile walls and eventually concrete walls—floodwalls—were built on top of existing small levees to increase their height and protect the ever-sinking city.

With the levees and floodwalls growing higher and the land subsiding lower, the city became a crater surrounded by walls that kept out the rising river or the rising tides. The only dry place was in the middle, thanks to the ditches and canals draining the city. By 2005 there were 172 miles of drainage canals, and thousands of miles of storm sewer lines, along with vast pumps constantly shoving water up and over the encircling walls and into canals to flow toward the lake. New Orleans resident and historian Douglas Brinkley described the city as "no more stable than a delicate saucer floating in a bowl of water."[23]

As cobbled together as this infrastructure was, the entities responsible for it were even more so. Each unit of government, from federal to local, had mixed-up and multilayered roles and responsibilities. Metropolitan New Orleans had a half dozen levee districts, each one a large, well-financed organization whose essential role was keeping its particular area of the city dry. These districts, like Nimrod's Mississippi Levee Board hundreds of miles upstream, had broad powers to collect taxes, sell bonds to finance flood-control projects, and administer a wide range of activities. In fact, these activities could and did extend beyond simply flood control to include land development. By 2005, one of the larger districts—the Orleans Levee District—had an annual operating budget of $40 million and over three hundred employees. It was in charge of maintaining just over 100 miles of hurricane levees with 127 floodgates as well as 28 miles of Mississippi River levees with 76 floodgates.

The Corps of Engineers was also an integral part of the flood control system in New Orleans. The Corps had been building levees and strengthening existing levees along the lower Mississippi River since 1928. The agency also began construction on its vast series of bypasses and floodways to divert flows out of the Mississippi River. But by the mid-twentieth century, the Corps began to realize that floods in New Orleans were as likely to come from the lakes due to a storm surge as to come from the river due to a flood; the flood of 1849 was the last time the Mississippi River levee was breached within New Orleans city limits.

To address the threat of hurricane storm surges, in the 1960s, Congress authorized the Corps to construct protection along Lake Pontchartrain. This work was to include barriers along the lake itself to defend against storm surges, but it would also protect the primary canals going from the lake into the city. Under normal conditions these canals work as drains to take seeping water away from the city to the lake. During a hurricane, they potentially reverse and serve as a funneled pathway for a storm surge to flow past the lakefront barrier and deep into the city. To address this potential, the Corps proposed using floodgates at the mouth of each canal where it enters the lake. The gates would close during a hurricane and keep the potential surge from entering the canals and funneling into the city. Floodgates at the mouths of canals would be, according to the Corps, effective and the least costly option.

The levee districts and the Sewerage and Water Board, which managed the drainage system, also had a say in the decision because they were required by Congress to pay for some of the project costs. The local boards and districts didn't like the floodgate approach proposed by the Corps. If the gates were closed, the accumulating rain and groundwater that normally would be pumped into the canals would have nowhere to flow. Keeping the floodgates closed for any period of time would fill the canals with water that then backed up into the city, causing more water backup in drainage ditches throughout the city and returning the city to its natural puddle-like condition. Local districts argued that instead of floodgates, the canals should be left open at the lake—leaving them susceptible to storm surges—but levees and floodwalls along the canals should be flood-proofed against these surges by being made higher and stronger. Higher floodwalls would allow a storm surge to enter the open-ended canals without overtopping. The districts would benefit further because the increased height of levees and floodwalls would help with their chronic interior drainage problems. Floodgates addressed only the issue of flooding due to storm surges; higher floodwalls would help with both flood control and interior drainage.

At that point, federalism raised its complex head: the Corps generally considered interior drainage to be a local issue and therefore not fundable by the federal government. So the local Sewerage and Water Board maintained internal drainage—all the pumps and ditches and canals—while the levee districts maintained the levees and floodwalls along the canals. As long as the levees along the canals were tied only to internal drainage, the levee districts had to pay all the costs of construction and maintenance. But by tying the floodwalls along canals to hurricane protection, the levee districts could potentially get the federal government to pay for a large portion of improvements to local interior drainage infrastructure. With the Corps equivocal on whether floodgates or floodwalls were chosen, so long as it was the least cost to the Corps, the levee districts lobbied and convinced Congress to authorize this alternative approach.[24]

From the perspectives of engineering and risk avoidance, these approaches were fundamentally different. The Corps would have focused its attention and resources on ensuring the integrity and functionality of three critical floodgates—effectively hydrological strongholds. The floodwall approach, in contrast, would require a strong infrastructure for the entire length of the levees and floodwalls along the canal system; the soundness of this flood control system would rely on the strength of miles and miles of levees and flood-walls. The adage about a chain being only as strong as its weakest link is appropriate in thinking about using a levee and floodwall sys-tem to protect a sub-sea-level city; a single weak point undermines the entire system.

Federalism relies on the local government doing its part—and on the central government not doing more than is needed. When infrastructure is involved, this means the federal government—the Corps—does some of the work but then hands off significant roles and responsibilities to local groups. One such case is Peter Nimrod's maintenance of the levees at Greenville. In New Orleans, the hand-off was complex and problematic. First were the half dozen levee districts, each with its own board of directors, spending priorities, and habits. Second was the involvement of various other government units, such as the separate sewerage board in charge of pumps in the canals. Adding to these layers, the Louisiana Department of Trans-portation had the role of approving activities that might compro-mise the levees, as well as directing training for levee district board members and inspectors. What's more, the local government units meant to focus solely on flood control and the levees had a surprisingly undisciplined attention span. The Orleans Levee District, for instance, used some of its revenue to build parks, marinas, an airport, and a dock for casino boats. In 1995 the district published a brochure that read, "We protect against hurricanes, floods, and boredom."[25]

Another problem was that risky future development was an essential element of the agreement to build levees along the lakes and canals. To justify the entire project when authorized and appropriated, the Corps and levee districts had factored projected new development into the cost-benefit analysis; fully 79 percent of the project's financial benefits were to come from future development rather than existing development. Jefferson Parish added 47,000 housing units, and Orleans Parish added 29,000 units in what had previously been swampland but was rapidly turning into drained, leveed, but subsiding suburban neighborhoods.[26]

After completing the construction work for its part of the project, the Corps had handed off most of the levees to the local districts, which were then expected to take up the responsibilities of operation and maintenance. But some levee districts refused to take over those responsibilities because they were unclear when they were to do so or because they were willing to take over only minimal roles of maintenance, such as grass and weed control. In general, the local districts claimed that "major" problems remained the responsibility of the Corps, while "minor" problems were the responsibility of the districts—a casual distinction at best. The conflicting or at least inconsistent roles of the different agencies could be seen in the elevation of the levees themselves; their heights varied depending on who was in charge. At one pumping station where jurisdictions had (opaquely) changed hands, a concrete wall was connected to an earthen levee that had a completely different elevation. The elevation of levees maintained by Peter Nimrod's Mississippi Levee Board is ensured with laser-like precision, but New Orleans had at least five different elevations of different parts of the levee system protecting the canals, and these variations likely corresponded to whatever agency was responsible for that section of levee.[27]

On August 29, 2005, the weaknesses in the New Orleans flood control system were laid bare as Katrina worked its way across the region. Gaps and inconsistencies in the floodwalls and levees created weak points through which water first came into the city. The storm made landfall just east of New Orleans at 6:00 a.m. At 6:30 the floodwall along the 17th Street Canal breached, sending water into Jefferson Parish just west of downtown. By 9:00 a.m. the Lower Ninth Ward had over 6 feet of water; later that morning a storm surge sent water into St. Bernard Parish, where water levels eventually reached 10 to 15 feet and covered 95 percent of the parish. The Industrial Canal floodwalls breached after one more hour, sending water into the Lower Ninth Ward through a gap large enough for an entire barge to be sucked through it and deposited in the neighborhood. When the London Avenue Canal levees breached later that afternoon, water poured into broad areas throughout New Orleans. It kept pouring through breaches in levees and floodwalls for two days, until noon on Wednesday, covering large swaths of the city with water. Only the old part of the city, sitting on the natural rise along the river, remained dry.

Federalism created the backdrop for flood control infrastructure that failed during Katrina, and federalism also drove part of the chaos of disaster response. Much went wrong before and after Katrina, but at the local level much also went right. In the two days before Hurricane Katrina, over a million people were evacuated from New Orleans—about 85 percent of the population—a significant achievement for a geographically isolated city. Following Katrina, search and rescue efforts were dominated by local and state organizations. The Louisiana Department of Wildlife and Fisheries, composed largely of personnel familiar with the affected area, used more than two hundred flat-bottomed boats to scour the city and rescue people from the rising waters. The locally placed Coast Guard likewise used 4,000 people, close to 40 aircraft, and dozens of small boats to conduct an outpouring of relief works.[28]

But much also went terribly wrong, particularly at the federal level. Response to Katrina was heavily influenced by the shadow of the terrorist attack of September 11, 2001. Before 9/11, disaster relief and FEMA reported directly to the White House, forming a direct chain of command from president to actions on the ground. Following 9/11, however, the administrative home of disaster relief was shifted to the Department of Homeland Security. This decision reflected the assumption that the most ominous disasters of the future likely would be terrorist attacks. Thus an apparently simple bureaucratic move by the federal government made natural disaster response the stepchild of homeland security.[29]

Federal disaster response was mortally slow, in part because of this reorganization. Federal officials were slow to appreciate the magnitude of the catastrophe. Over 24 hours after 20,000 people had fled to the Morial Convention Center, where they were trapped without food or water, Secretary of DHS Michael Chertoff stated on National Public Radio that "Actually, I have not heard a report of thousands of people in the convention center who don't have food and water." Even after federal responders recognized the scale and severity of the situation, FEMA's response was governed by the new 400-page *National Response Plan*, a manual prepared in the wake of 9/11, which became a chokepoint for federal decision making in the rapidly evolving events following Katrina.[30]

The use of National Guard troops was an obscure but critical battle line drawn by federalism amid the fetid waters of Katrina. In the late twentieth century the National Guard was typically used for natural disaster recovery, but its involvement was always in response to explicit requests by governors of affected states. This practice represents an important distinction between the National Guard and the regular Army. The National Guard is intended to be the military presence within U.S. boundaries and under the command of the affected state's governor. Federal troops—the regular Army—cannot be used within the United States unless the state government is unwilling or unable to suppress violence that is in opposition to the U.S. Constitution.

A president can also federalize National Guard troops, effectively putting a state militia under federal command; this would be much the same as sending in federal troops to regulate state behavior. While seemingly outdated, this practice is a militaristic version of federalism: a state check on federal power. In 1957, against the governor's will, Eisenhower used the Army's 101st Airborne Division and also federalized the Arkansas National Guard to enforce school desegregation in Little Rock. It was one of the very few times that such actions have been taken. Sending federal troops to enforce law and order against the wishes of a state governor, or federalizing a state's National Guard against the governor's wishes, raises the ire of federalists perhaps more than any other potential action a president could take—especially in a southern state.[31]

Against the backdrop of a flooded New Orleans, Louisiana Governor Kathleen Blanco requested federal troops to assist in disaster recovery so that Louisiana National Guard troops could focus on law enforcement in the city; this way, federal troops would be doing humanitarian work only while National Guard troops would be conducting actions of a more militaristic nature. The states affected by Katrina did not have full use of their own National Guard troops because in late summer of 2005, when Katrina struck New Orleans, about 40 percent of the National Guard units from Louisiana and Mississippi were in Iraq for the war on terrorism. The U.S. Northern Command, a unit of the federal military created after September 11, 2001, to protect the continental United States, moved to the Gulf Coast to support FEMA. But stories of widespread looting and murders coming out of New Orleans led the Bush administration to question whether local law enforcement and the National Guard, under local command, were really up to the task. Yet federalizing National Guard troops or sending in federal troops would require the governor's permission unless an actual insurrection was in progress.

Despite the continuous violence and chaos, looting is not the same thing as an insurrection.[32]

On Friday night, four and a half days after the first levee breach, Louisiana Governor Blanco received a fax from the White House. With her signature, that document would federalize the National Guard troops under her command. Early the next morning, as Bush was preparing a Rose Garden announcement of his decision to federalize the Louisiana National Guard and other rescue functions, Governor Blanco refused to sign. Though his stance was less publicized, Mississippi Governor Haley Barbour—a staunch Republican ally of the president—also refused a request from the White House to turn over his National Guard troops to the federal government. Rather than call in the federal troops, both Louisiana and Mississippi stuck to their own local resources as well as relying on the 20,000 National Guard troops sent in from other states in the region.[33]

It seems strange to manage a government function as critical as flood control through a system as complicated, redundant, and seemingly undisciplined as federalism. It's unlikely that anyone designing a system from scratch would have had Kent Parrish and Peter Nimrod overseeing many of the same levees from completely different levels of government. The inefficient and complicated effect of federalism for flood control, and almost all other government functions, seems to be something that the framers of the Constitution fully expected in their design of government itself.

We can better understand that element of the design by viewing the American government as something intended to be an ongoing experiment rather than an optimized system. That is, the framers constructed government with a willingness to try something, test the results, keep what worked, and reject what did not. By rejecting the Articles of Confederation and reconstituting the entire government in 1787 under the Constitution, they set the most extreme example; but even as they were designing this new government, they embraced experimentation as a core part of their vision. The *Federalist Papers*, which argued the case for the states to ratify the new Constitution, use the word *experiment* more than forty times, but *democracy* less than a dozen. Most of the early presidents described the government frequently as an ongoing experiment. The purpose of such an approach was captured well by Andrew Jackson's 1829 State of the Union address: "Our system of government was by its framers deemed an experiment, and they therefore consistently provided a mode of remedying its defects."[34]

If we accept that government is an ongoing experiment in which we propose and test different methods of accomplishing different functions, then the seemingly chaotic structure of federalism starts to emerge as something far more useful. Changing agencies or entire governments at the national level is extremely disruptive; adapting smaller, more local governments or agencies is far more realistic. That is, federalism provides the opportunity to conduct a series of smaller, simultaneously occurring experiments. Failure or success from each experiment can be used to inform those in other states or local governments, and failure of one of these smaller experiments does not jeopardize the entire organization.

The trial-and-error method of Mississippi and Louisiana's earliest locally managed levee districts led to the creation of similar but slightly different districts in Arkansas. The idea eventually was used in Illinois and across the continent in California. Each state and each levee district could adopt what it liked from others and adjust it for their own peculiarities, whether those peculiarities were hydrologic or economic. This rapid adaptability is what led Supreme Court Justice Louis Brandeis to refer to states as "laboratories" of policy where ideas could be tried at a limited scale, and their merits weighed based on performance, before potentially being implemented nationally. Such adaptation is far more difficult when policies or approaches are nationalized. The building of mammoth flood control infrastructure was largely unstoppable through much of the twentieth century, just as curbing federal spending on flood insurance is proving almost impossible. Momentum at the national scale precludes making changes; too much is invested in the status quo by too many people.[35]

Besides its adaptability, federalism also appealed to the founders' aversion to vesting too much authority in a small number of people in just one agency. When flood control was centralized in the Corps of Engineers during the Great Depression, the scale of work done was staggering in comparison to what had been done until then. Just as important, the function was controlled and operated by a small set of people in the Corps of Engineers; this arrangement contrasts markedly with the highly diffuse approaches of local or state efforts such as the dozens of levee districts along any major river. Making flood control a federal responsibility led to centralized control of entire river systems. The Corps of Engineers, with congressional authorization, decides how much water will be released from Missouri River dams or when to activate bypass channels on the Mississippi River. Farmers and local entities must adjust their practices and respond each year to changes in management decisions by the Corps in much the same way that they must adjust to changes in the weather.

And this is the crux of federalism: our view of how active local versus federal government should be in controlling or responding to floods reflects our view of human nature itself. Underlying the call for a strong federal role in flood control is the vision that river managers and engineers have the foresight, expertise, understanding, and public interest to make holistic and integrative decisions to alleviate the threat of floods across the nation; that the best and the brightest people working in flood control are consolidated in a singular federal agency. The alternative vision is that rivers and floods should be managed among disparate personnel, institutions, agencies, states, and local districts. Resistance to nationalized flood control is based on an ideological vision that singular centers of decision making are apt to become victims of bureaucracy, or perhaps simply implementers of decades of previous decisions.

If we accept that there is a cadre of visionary and superior river engineers and managers whose abilities can be harnessed to alleviate floods or enable navigation on rivers, then we should ensure that those river masters have the resources and authority for developing and implementing massive projects and policies. If instead we are inherently skeptical of any river managers, regardless of their capabilities or purity of motives, then it is better to constrain them by diffusing authority: give certain functions to a federal agency, but retain others for the states and even local, function-specific districts like the Mississippi Levee Board.

This, then, is why Americans have developed a complex and somewhat unintuitive approach to flood control, and river management in general: because of our federalist government. And this federalist approach to government is based on our ever-evolving, pendulum-swinging perspective on how much power and influence should be vested in a few people—or in many people. In the end, how we as a nation have approached managing rivers and floods is a manifestation of our ideology; or as James Madison stated in *Federalist* No. 51, "But what is government itself, but the greatest of all reflections on human nature."

PART TWO

SOVEREIGNTY AND PROPERTY

CHAPTER 5

Water Wars

The Klamath River starts in the high desert of central Oregon and cuts through scrubby rangelands, dusty canyons, and the redwood-lined mountain ranges of Northern California, flowing eventually into the Pacific Ocean just north of Arcata. The river links Oregon to California and ranchers to salmon. It is also contested territory in an ongoing water war.

Water wars could often more accurately be called hydro-political skirmishes, but they can feel as serious as a war for those living in the affected regions. Most of these water wars have been waged in the West, where surrounding aridity and recurrent droughts regularly ignite hydrological conflict. Battle lines have perhaps most frequently been drawn along the Sacramento and San Joaquin Rivers of California's Central Valley and the Colorado River, deep in the dry Southwest. But at the start of the twenty-first century, tensions over water have been rising almost everywhere—between Texas and Oklahoma, between the Great Lakes states, and even between Georgia, Alabama, and Florida, states that occupy one of the wettest regions of the world. The battle for use of the Klamath epitomizes these skirmishes, for it includes many quintessential elements of water wars: fish, farmers, Native Americans, and the struggle to maintain highly water-dependent lifestyles in the high desert. By 2013, Klamath River water wars had been pitting upper-river farmers and ranchers against their downstream Native American neighbors for decades, and another skirmish was brewing.

The Klamath watershed is shaped like an hourglass tipped on its side; a narrow canyon separates expansive upper and lower basins. A series of high dams sits in the canyon, starkly dividing the two basins. The upper basin is arid, flat, and bordered by a distant ring of forested mountains. The lower basin is steep, with ragged mountains that creep to the edge of the river, and slightly more humid—at least by western standards.

Down in the tribal lands, despite flowing hundreds of miles, the Klamath is no bigger than it is in the upper basin. Rivers should get bigger as they flow downstream: like arteries, they gather the flows of the many tributaries that feed them, swelling along with the area of land and runoff they accumulate. But the Klamath is more like a vein, distributing its water out onto the landscape for irrigation through the ubiquitous water diversions along its course. Most rivers of the West have a hydrological fate similar to that of the Klamath.

In the arid West, it is a constant battle to grow anything. Farmers and ranchers can scratch out a living if, and only if, they can get water. They live and die by building "diversion dams" that transfer water away from a natural channel and out into the landscape through a series of progressively smaller canals and ditches that sprawl into various fields. Irrigation works can be as simple as a piece of plywood carefully wedged to spew water over a field or as complex as a massive concrete dam shifting water into a cross-state canal. Westerners use these diversions to move water around, and moving water around is critical. In fact, farmers here are often simply called irrigators because irrigation is the central activity of farming in the West.

The spokesman for some of these irrigating farmers and ranchers in the Klamath watershed region is Tom Mallams. He was elected commissioner of Klamath County, Oregon, during the rise of the Tea Party movement, which resonated in the independent-minded western lands as strongly as anywhere in the United States. Mallams is an eminently polite farmer-politician, soft-spoken but clear-headed and razor sharp. When I meet him he is rail thin and deeply tanned from years in the high desert, dressed in a dark suit and red tie. His office is immaculately clean and organized. Mallams knows water, and having lived and farmed through sequences of droughts, he's all too aware of what the loss of water would mean for farmers—irrigators— like him and the county he represents.

A few hundred miles downstream from Tom's office, past the canyon that funnels the river into the lower basin and on through the craggy mounts of the Coast Range near the mouth of the Klamath, Leaf Hillman lives on the lands of the Karuk Tribe. At first glance, he and Mallams don't seem to have much in common. Hillman is Native American. He's short and stocky, and he has a long ponytail of black hair; the day I meet him, he has worn shorts and a tank top to work. Whereas Mallams's county office was in a renovated building with well-manicured bushes along its steps and stairs, the Karuk Tribe Natural Resources office where Hillman works is drab and undersized. The tribal health care center is in the same building, which is perched in a narrow canyon next to a highway. Despite their many differences, Hillman and Mallams are bound together by the Klamath River.

Leaf Hillman knows Tom Mallams, like he knows many of the farmers and ranchers in the upper basin. For his part, Mallams knows Hillman and the tribal members throughout the basin. Of course they know each other. As Hillman says remorsefully of the upstream farmers, "They were set to be my enemies before either of us was born." And of the things that make them "predetermined enemies," water is central.

Both Hillman and Mallams have spent their lives facing the realities of water scarcity. Mallams knows firsthand how hard it is to scratch out a living on these dusty landscapes without sufficient water. But farming is not just a job to him—"it's a lifestyle; a culture."

Hillman and the tribes are all too aware of what is at stake for the farmers: "If they don't have certainty in water, then they have no stability in agriculture, and then they have social instability in their people and their place. The irrigators need social stability." Hillman can see a failed hydrologic future of atrophying farm towns, shut-tered businesses, decaying schoolhouses, weedy fields, and kids leav-ing home to live somewhere, anywhere, else—a farming version of the desolate tribal lands that still pepper the West. "If they are going to have stability in their communities, they need to have stability in their water."

Hillman empathizes with the farmers the way only a fellow water-starved Westerner could, largely because he has watched as tribal water was winnowed away through decades of water diversions by upstream farmers. As the diversions reduced the flow in the Klam-ath, the number of salmon in the downstream river plummeted. As the salmon declined, so did a revenue source for the lower basin tribes, making their livelihoods—their lifestyle and their culture—ever harder to sustain.

Yet despite this intense history, the ranchers and tribes don't wish each other ill. They have a common respect for each other—as Hill-man described the upstream farmers and Mallams, "They are hard-working sons-a-bitches. They just want to live their life and be left alone to work and be with their families. We have a lot more in com-mon than we think." Mallams, likewise, counts many tribal members throughout the basin as friends. But water in the West is a zero-sum game. The tribes can't keep enough water in the river to sustain the salmon if the farmers upstream divert all the water they need for crops and cattle. Someone has to lose. Furthermore, someone has to choose the winners and losers.

As I left Mallams's office, I passed an enormous steel bucket standing almost 15 feet high, emblazoned with the name *Oregon*, along with the names of several other western states: Utah, Idaho, and Nevada. The bucket is a tribute to the "bucket brigade," an event that occurred in 2001 during the last major drought to hit the Klamath basin. In that earlier drought, the federal government had not allowed water to be diverted into irrigation canals. Instead, water was left in the streams to protect the endangered species living in the lakes and rivers of the basin: two species of suckers in the upper basin, and coho salmon in the lower. In response, farmers and ranchers from other western states, who resented what they saw as federal overreach, formed a line through the town of Klamath Falls and passed buckets of water by hand from the river into the irrigation canals.[1]

The 2001 bucket brigade was a twenty-first-century sagebrush rebellion, a natural extension of the western tradition of rebelling against perceived federal government overreach of authority over land, water, and other natural resources. The original sagebrush rebellion erupted in 1979, when rebels in Utah bulldozed a barrier on federal land and claimed the land rightfully belonged to the county and state of Utah. In Nevada, where the federal government controls 87 percent of the state, the legislature passed an act requiring all federal lands in the state to be reevaluated and potentially made state lands instead. In 1980, when then presidential candidate Ronald Reagan was asked about the sagebrush rebellion, he declared, "Count me in."[2]

Sagebrush rebellions did not arise from the western farmers' quest to eradicate government, but rather from their quest to ensure that the states and not the federal government would have final say in deciding how to use natural resources like land, timber, and water. The sagebrush rebellions were all about *sovereignty*.

Tom Mallams and the farmers of the Klamath pointed to the State of Oregon as the sovereign entity over water—and as sovereign, the state had granted these farmers water as their *property*. And they are right. Leaf Hillman, his Karuk Tribe, and other tribes in the Klamath had a different perspective on who makes decisions about water: their view derived from over a century and a half of being on the losing end of decision making. Hillman noted that, at least in theory, the United States recognizes the sovereignty of the tribes; and as sovereign nations, the tribes have as much right or even more than the state

of Oregon to the water from the Klamath River that their reservations have historically received. He and the tribes are also right. Both
groups are right, depending on who is in charge. Who is sovereign?

Think of how we divide land: we draw lines on maps. Some lines
follow nature, tracing the wiggling (meandering) path of a river or a
mountain range. Others follow politics, tracing the subjective precision of latitude and longitude. Once those lines are drawn, we revere
them. We bound territory with these arbitrary lines on a map, and we
recognize the authority of a person, or a group of people, within that
territory. This is sovereignty—a geographically bounded territory
with some ultimate authority ruling over that territory. On a smaller
scale, lines on maps can also designate property: on one side of the
line is my property, and on the other side of the line is your property.
The historian Bill Cronon simplifies the concept like this: Property is
you vs. me; sovereignty is us vs. them.[3]

We are typically comfortable with the ideas of property and sovereignty as ways to divide land. Fences dividing fields or lines dividing
a map; both are intuitive. Dividing water is not so intuitive. When
drawing lines in water, we are confronted with an unfortunate characteristic of water: water flows, often across the boundaries between
individual properties and sovereign groups. This doesn't stop sovereign authorities like the Karuk Tribe and the State of Oregon, or
property owners like the Klamath farmers, from trying to divide it
anyway. To understand conflict over water requires thinking of rivers
through these dual concepts of property and sovereignty. An additional conundrum is that while the U.S. Constitution recognizes
the property rights of its citizens, it also recognizes the authority—
literally, the sovereignty—of several different groups: the federal government, the states, and the tribes. Water wars are not caused by
scarcity alone. They are about scarcity amid the realization of these
ideas of property and sovereignty as applied to water.

To understand the western method of treating water as property, we need to back up 150 years to Sutter's Mill, California, in the chaotic aftermath of the discovery of gold when states in the arid West were just beginning to be created. In January 1848, just days before California officially was recognized as a U.S. territory, carpenter James Marshall found traces of gold just downstream from a sawmill—Sutter's Mill—he was building on the South Fork of the American River in California. Marshall's discovery not only kicked off the gold rush but also shaped the development of western water law.

Water law is all about how water can be used—whether it can be impounded, diverted, and otherwise moved out of and away from a natural waterway such as a stream. In the East, water law was, and is, based on common law—the set of laws and practices that the colonists brought with them from England. Under common law, a riverfront or "riparian" property owner has the right to use water from that stream, but only an amount that would constitute "reasonable use" of the flow of the stream. Upstream riparian landowners are expected to leave enough water for downstream riparian landowners to enjoy the same right to reasonable use. Location is critical in the East: the water right is based on ownership of the riparian land. While this application of law is often attributed to the East being water rich, as important is the condition of the territory where the law was being applied: when water law was being established and developed in the eastern United States, the vast majority of land was privately owned, just as land was in England. It made sense to use English laws because land ownership was almost identical to what it was in England. Because of its emphasis on location, eastern water law has become known as the "riparian doctrine," and states east of Kansas City invariably use some form of the riparian doctrine.[4]

California was the first territory settled in the West, and it initially adopted the customs, like the riparian doctrine, that settlers had brought from back East. The riparian doctrine is built on several assumptions: that a stream will furnish plenty of water to go around; that people might reasonably want to irrigate only the land immediately next to a stream; and that the land being irrigated would be privately owned. But just about every assumption undergirding the riparian doctrine broke down during the frantic building of mining camps in the Sierra Nevada in the aftermath of Marshall's discovery at Sutter's Mill.[5]

Mining in California first took place in riverbeds, but miners quickly discovered that much more gold was in the banks of rivers as well as in veins extending far out into the surrounding hills. To mine riverbanks and hillsides, miners diverted water out of streams and rivers and into hand-dug ditches or wooden flumes. Allowing water to be diverted and moved so far from riparian land separated water ownership from land ownership, which was the first ingredient in considering water itself as property.

The second ingredient was the timing of the gold rush. During the initial rush, California was a sparsely settled federal territory. Even after it became a state in 1850, most of the lands being mined were in the public domain—they were owned by the federal government rather than by private individuals or the state. According to custom, it was perfectly legal and even encouraged for private citizens to make a living off these public lands. In fact, before the discovery of gold, Sutter's Mill was intended to cut wood that was timbered from public forested lands in the same way that timber companies during the 1900s and 2000s cut trees from Forest Service—public—lands. Because most of the lands in the West were in the public domain, water rights could not be tied to land ownership; it was impossible to use the eastern riparian rights system.[6]

Both of the basic mining actions involving water—diverting water to non-riparian lands and using water from streams on public lands for private profit—divorced water rights from land rights. This decoupling was the essential conceptual move for the development of western water law. Riparian land was no longer a prerequisite for asserting water rights—water itself had become property.

The next two developments in western water law are best thought about by envisioning the pioneers of the mid-nineteenth century—whether the "forty-niners" of the gold rush in California or the "Sooners" of the Oklahoma land rush—frantically dashing into unknown and unsettled regions in hopes of making their fortunes. In both cases, most of the areas being settled were territories or infant states where the federal government owned much, or even most, of the land. These public lands were available for private use to the first settlers to "stake a claim," whether for a Sierra Nevada gold mine or an Oklahoma homestead. Natural resources on public lands were free for the taking.

But the claim couldn't be made and then abandoned; the land and the resource had to be put to use. In California gold-mining areas, the mine had to be continuously worked to own the claim; in the farms of the Great Plains, the land had to be continuously farmed to own the claim (thus Laura Ingalls Wilder's family had to live in their miserable "claim shanty" to sustain residence on their land claim). In the case of a water claim, the water had to be diverted from a stream and put to continuous "beneficial use," whether for mining gold or irrigating agriculture. This requirement was in contrast to the eastern riparian doctrine, which stipulated that a riparian landowner always retains the right to water based solely on land ownership, whether or not the water is ever put to use. In the western United States, water rights were dependent on staking a claim and then putting the water to use.

A key concern in the emerging "appropriation doctrine" of the arid, booming West was the seniority of the claim, or when it had been made, compared to other claims on the water. Though anyone could come to the same stream years or decades later and claim water, the senior water rights were satisfied in full first, and any remaining water went to the junior rights-holders in chronological order of their claims. These senior claims were sustained as long as the water was put to beneficial use. If a farmer didn't use all the water he had claimed, he would permanently forfeit his unused portion to the junior water claimants. This appropriation doctrine for establishing property rights for water came to be known by two phrases: "first in time, first in right" and "use it or lose it."

When the territories became states they preserved the doctrine, for they found it an effective way of solving some critical issues of farming and irrigating in the West. Because the western landscape is arid, and few streams are available to divert, farmers had to move water great distances to get it to their property. The first two irrigation canals in Greely, Colorado, were 16 and 36 miles long. Wyoming's Big Horn basin required an irrigation canal almost forty miles long. Even the early Mormon settlements of Utah in the mid-nineteenth century, set amid the peculiar stream-rich areas of the Wasatch Mountains, needed canals averaging almost four miles long.

Any infrastructure for farming is an investment. But the infrastructure needed for western irrigation, rudimentary as it was, called for significant financial risk in a region where capital was scarce. For farmers to invest in the dams, canals, and ditches necessary to put the water—and the land—to use, they needed a future guarantee of their right to use the water they'd gone to the trouble of diverting. This was the elegance of the appropriation doctrine. For early farmers considering whether to build the first dams and canals, making water use a property right based on seniority offered the level of certainty they needed. As long as there was enough water to go around, all of the water claims could be filled. But in drought years, those with the most senior claims—senior appropriators—were assured of getting all of their water while junior appropriators were forced to wait in a hydrological queue, hoping there would be enough to go around.[7]

Each of the western states developed its own particular approach to property rights, but all of them were modeled on the appropriation doctrine. Importantly, states decided how water was to be treated within their borders independently of the federal government. Just like states already existing in the East, western states had sovereign power over the water within their borders, and they developed their property rights systems independently of each other.

It was generally accepted that states were sovereign for dividing water within their borders—for determining the meaning of property within their borders. More contentious was the question of who had the authority to divide water that crossed borders.

Lee's Ferry may be the single most important location on any river in the United States. It is where we must pivot from thinking about water as property to thinking about water and sovereignty.

Anyone moving across the desert Southwest, whether Spanish priests in the eighteenth century or Mormon settlers in the nineteenth century, found the Colorado River canyon mostly uncrossable from Hite, Utah, to the Arizona–Nevada border 450 miles downstream. Like the Klamath River, the Colorado watershed has an upper and lower basin, divided by a canyon—in this case, the Grand Canyon. At its head sits Lee's Ferry, on a topographic oddity of relatively gradual slopes down to the river that created one of the only river crossings in this otherwise impenetrably long and deep canyon.[8]

This peculiar point on the map demarcates two groups of states: the upper basin states—Utah, Colorado, New Mexico, and Wyoming—and the lower basin states—Nevada, California, and Arizona. In the early twentieth century, as the population of the West grew dramati-

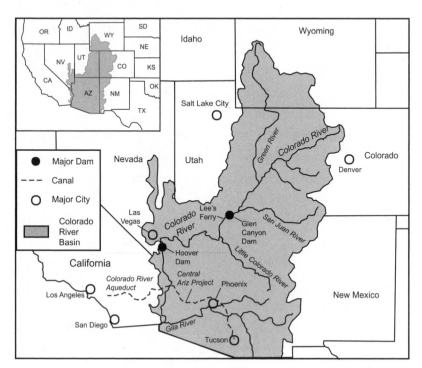

The Colorado River Basin.

cally and states increasingly disputed each other's claims to the limited supply of water, it was far from clear who had the right to the water in the Colorado River. Interstate water disputes like that over the use of the Colorado River revolve around sovereignty and are resolved through interstate treaties.

Sovereignty is a slippery concept. We can trace our concept of it back to 1648 and the Treaty of Westphalia, the agreement that ended the Thirty Years' War in Europe. While the treaty is most immediately notable for ending a long and bloody conflict, its lasting legacy has been establishing the way governments relate to each other: as coexisting, sovereign states. Since Westphalia, groups of people are not identified as sovereign unless they occupy a place that is governed by a recognizable authority, and the bounds of that place are identified and recognized by others.

This notion of sovereignty is why modern geopolitics depends so heavily on borders. Groups of people are not recognized unless they are confined to and defined by a place. Native American tribes, for example, are nations of people—each tribal nation shares common language and common descent. But they didn't have distinct, bounded territories, so there was no precedent for how the British colonists or early U.S. citizens should treat the tribes in modern governance, where place—territory—was part of being sovereign. This situation made it easier for first the colonists and then the U.S. government to claim sovereignty over land that Native American nations had been using. After the eventual establishment of tribal reservations, the U.S. government had a familiar way to recognize the tribes as sovereigns. Now tribes are inextricably connected to their reservations in the eyes of the U.S. government, whether the tribal nations are located in their traditional geographic regions or not. Similarly, the nation of Israel was not recognized as the sovereign State of Israel until the Zionists established a bounded territory in Palestine. The reason sovereignty is so critical to establish is that, since the Treaty of Westphalia, other sovereigns have recognized that an ultimate authority governs a particular territory.[9]

In the United States, as Leaf Hillman noted along the banks of the Klamath, the federal government, state governments, and Native American tribes are all sovereign, and yet their territories all overlap. In water wars, state governments want to have final say over how the water within their borders can be used. But rivers recognize the geography of topography rather than that of politics, and therefore water must be divided and assigned a sovereign when it crosses borders. Lee's Ferry is important because it became one of these arbitrary political boundaries on the Colorado River, the most important river of the Southwest.

The challenge at the turn of the twentieth century for water in the West was to resolve who would make the rules for divvying up water and what those rules would be. The question of who would make the rules for dividing interstate rivers was quickly answered: the federal government, through the Supreme Court. The Supreme Court has been regularly called on to resolve interstate water conflicts since a 1902 conflict between Kansas and Colorado. In *Kansas v. Colorado*, the state of Colorado took the approach that when it came to water, states within the United States should be treated as individual, sovereign governments: "Colorado occupies toward the State of Kansas the same position that foreign States occupy toward each other." The U.S. Supreme Court did not accept this absolute sovereignty argument, instead deciding that states would be treated as equals, and the Supreme Court would be responsible for establishing justice between the states. In essence, the Court said that it was the sovereign authority over water within the bounds of the United States.

Although the Supreme Court declared it would be the decider, its basis for making those decisions was completely unclear. During that era, many dramatically different and competing philosophies of interstate water rights emerged.

On one hand was the Harmon Doctrine, named for an opinion written in 1895 by the U.S. Attorney General Judson Harmon. In

1894 Mexico had made a series of complaints against the United States about its heavy diversions of water from the Rio Grande before the river reached Mexico. Before entering negotiations with Mexico, the Secretary of State, Richard Olney, needed some legal basis for justifying U.S. water rights in the area. Olney asked his attorney general, Harmon, for some principles of international law for discerning how to proceed. Harmon's opinion was a full-throated patriotic argument in support of U.S. water rights; he insisted that the United States owed Mexico nothing: "The fundamental principle of international law is the absolute sovereignty of every nation, as against all others, within its own territory." The Harmon Doctrine of absolute sovereignty obviously favored any upstream water user—which, in the case of the Rio Grande, was the United States.[10]

Though the Harmon Doctrine was never used as originally intended in negotiations with Mexico, many U.S. states used it in dealing with their own interstate conflicts. When upstream Colorado set out its case against downstream Kansas, Colorado combined the idea of state sovereignty with the Harmon Doctrine, asserting that "As a sovereign and independent State," Colorado had absolute sovereignty "over all things within her territory, including all bodies of water," and thus could not be obligated to deliver any water past its borders.[11]

At the other end of the spectrum was the philosophy that older, established water users should take priority over newer or undeveloped water users. Essentially, this scheme applies the prior appropriation doctrine for water allocation between users to water allocation between states. Even though it would have given Mexico greater right to water from the Rio Grande than the United States, this approach had substantial international and historical precedence.

Historically, in arid regions, the older, more economically and politically developed water-using societies typically were downstream relative to emerging water-using groups. These older water-using groups—Egypt on the Nile, Babylon on the Tigris-Euphrates, California on the Colorado—have generally seen upstream development as a threat to their water security, because new water users upstream might at some future time drain their water source. Downstream groups inevitably prefer rights based on historic use, like those outlined in the appropriation doctrine. For their part, upstream users tend to prefer rights based on contribution to river flow, like those of the riparian doctrine. This struggle was certainly obvious in how states approached dividing the water of the Colorado River: upper basin states, most notably Colorado, all made some sort of absolute sovereignty argument; downstream states, most notably California, all claimed rights deriving from existing historical use. Everyone expected that the Court would have final say on dividing up the Colorado River. After all, the states themselves seemed unlikely to reach consensus on how to divide the water.[12]

As Secretary of Commerce, Herbert Hoover was an unusual politician for the early twentieth century. He had effectively no political experience; after making a fortune as an international mining engineer, he then concentrated on delivering food to war-torn Belgium during World War I and became known as the Great Humanitarian for his leadership in that effort. He was also unusual for being from the West; when elected president in 1929, he was the first to come from a state west of the Mississippi River. In 1922, due to his experiences in the West as a geologist and mining engineer as well as his international reputation as a problem solver, Hoover was appointed the sole federal representative on—and elected the chair of—the Colorado River Commission. It was a new interstate commission, created by the seven states of the Colorado River basin to divide water rights to the Colorado without having to rely on the notoriously unpredictable Supreme Court.

With no real guidance from the Constitution or legislation from Congress, the Court had no precedent for deciding how to divide the water from the Colorado. One way to avoid leaving the decision up to the Court was for warring states to preemptively negotiate a compact, essentially a treaty between states of the United States. An interstate compact for the Colorado River basin states would put the decision in their hands: once passed, it would become federal law, binding the federal government—including the Supreme Court—and all other states. The downside was that a compact must be approved by each state legislature as well as by the U.S. Congress and thus would require negotiation and compromise.

As is typical in water wars, the new commission was obliged to contend with a slew of historical claims and arguments in the vein of the Harmon Doctrine while attempting to draft an agreement that would stand the test of time. California, the wealthiest and most downstream state, was using ever-increasing quantities of water for its sprawling cities in and around Los Angeles. Burgeoning agricultural communities in the southern California desert also needed water, and there was no indication that the growth would slow anytime soon. In contrast, most other states in the basin were poorly populated and had limited existing agricultural development, but they had a lot of land that eventually could be irrigated and used for agriculture. If the Harmon Doctrine of absolute sovereignty were applied, then upstream states could just divert whatever water they needed from the river as their cities and farms grew large enough to need it. If, however, some form of prior appropriation doctrine were used, California would get the windfall of water based on its established current water use in 1922, even though there was no infrastructure in place to put the water to use at the scale California thought possible.

The compact negotiations were necessary to resolve these conflicting pulls: upstream states wanted to make sure that California didn't take all the water while California wanted to get the pieces in place for large-scale—federal—water infrastructure. California, for instance, needed to have a massive flood control dam built to protect a growing agricultural region that had suffered a series of devastating floods; in 1905 the Colorado River had left its normal course and flowed 80 miles west, destroying thousands of acres. California thus wanted a flood control dam along with the ability to divert water out of the river. But before any dam on the river could be built, federal or not, California had to obtain the right to be diverting the amount of water it intended to. Similarly, for Arizona to make use of Colorado River water to irrigate fields and grow cities, it would need an enormous cross-state canal system, a scale of infrastructure well beyond what Arizona alone could finance. Upstream states likewise had limited ability to store any water, let alone convey that water with canals for cities or farms. All of these types of infrastructure were well outside the capabilities of the individual states, and representatives from these states knew they could never use their clout to get federal funding for the necessary infrastructure as long as the states were at loggerheads over who got how much water. The Colorado River Commission headed by Hoover provided a setting for states to begin resolving these issues—not by determining the allocation of funds, but by specifying the allocation of water to make future projects even possible.[13]

After the first six months of meetings and hearings ended in abortive failures, the Supreme Court forced the commission's hand. In June 1922 the Court finally released its decision on the conflict between Colorado and Wyoming for the water of the Laramie River. It said that in western states where the prior appropriation doctrine was used for dividing up intrastate water, the appropriation doctrine would also apply to interstate conflicts. Along with this finding, the Court definitively rejected the notion of absolute sovereignty proposed in the Harmon Doctrine. Upon hearing this decision, the upstream states were quick to realize that California was now fully in the driver's seat, and the only way they could protect water rights for their own future development was to secure them through an interstate compact.

Hoover had the commission sequester itself that November at Bishop's Lodge, a ranch in New Mexico. He chose the lodge because it was several hours away from Santa Fe by rough roads, thus limiting the potential for observers and interlopers. At the ranch, formal talks were held with only the delegate and advisors (an engineering adviser and a legal adviser) from each state; this approach minimized outside influence or lobbying during the actual talks and allowed the commissioners to negotiate concessions without immediately infuriating their state interests. These conditions, combined with Hoover's constant cajoling and needling and prodding, resulted in a compact that hinged on a great compromise.[14]

The first part of the compromise, and the negotiating breakthrough that made the eventual Colorado River Compact of 1922 possible, was at first a punt. The delegates initially were trying to divide the water seven ways—between the seven basin states—but could not. In a last-ditch effort to stop the whole discussion from collapsing, they arrived at the option of dividing the water first into the upper basin and a lower basin and then letting these subgroups of states continue to fight among themselves. Lee's Ferry was a convenient geographical point between states, roughly dividing the entire Colorado basin into upper states (Colorado, Wyoming, Utah, New Mexico) and lower states (California, Arizona, Nevada).

The second part of the compromise required the states to determine how much water was available to be divided. The U.S. Geological Survey had been measuring the flow of the Colorado since the late nineteenth century, and based on the data available in 1922, the average annual flow at Lee's Ferry was estimated at about 16.5 million acre-feet (MAF). Because one acre-foot equals the amount of water needed to cover an acre of land with a foot of water, about 330,000 gallons, 16.5 MAF adds up to roughly 5.3 trillion gallons. The 1922 compact said that the upper basin states would not let the 10-year total flow at Lee's Ferry fall below 75 MAF, effectively giving the lower basin states a guaranteed average of 7.5 MAF per year. From there it would depend on the flow in the river; in normal years, the lower basin would get 7.5, the upper basin would get 7.5, and then if sufficient water was available, the lower basin would get an additional 1.5 MAF. Moreover, the basins would split their obligation to Mexico.

Later acts and court decisions would further refine and specify water distribution within the particular states of each basin and formalize an obligation to Mexico of 1.5 MAF. The accumulated agreements, treaties, Court decisions, and legislation for divvying up the water of the Colorado—between the United States and Mexico, between the upper and lower basins, and between the states within each basin—are now generally known as the Law of the River.[15]

An important element of the 1922 compact, and of subsequent additions and refinements, was that everything grew out of the prior appropriation history of water use in arid regions. First, Mexico would be the most senior appropriator; all the states would ensure that Mexico's portion got filled, and the upper and lower basins would share that burden. Next in the hydrologic line was everyone— individuals living along the river—who was using Colorado River water at the time of the 1922 compact; these would be designated as "present perfected rights." For all intents and purposes they would be the most senior, inviolable water rights in the United States to the Colorado River system, the last water to be curtailed during the most severe of droughts. Once these senior water rights had been filled, based on the compact requirement to deliver 75 million acre-feet over a 10-year period at Lee's Ferry, the lower basin effectively would be the most senior user. Then, as long as the water supply was sufficient, the upper basin states would divide the remainder, which was expected to be 7.5 MAF.

The final compromise of the 1922 compact was a concession to California. Because California had the most political leverage, there was general recognition that the other states would support California's bid for a flood control dam in the lower river, downstream of the Grand Canyon. While California wanted that concession made explicit in the final 1922 compact, the other states refused, suggesting that all the states push for the dam after Congress had recognized the final compact.

Once the commissioners had signed the compact, it had to be ratified, first by their state legislators and then by the U.S. Congress. This step required a series of further political concessions. To convince legislatures of the upper basin states to agree to the delivery

Allocation of the Colorado River—The Law of the River

Upper Basin—7.5 million acre-feet per year (2.4 trillion gallons)

State	Acre-feet per Year
Colorado	3.86
Utah	1.71
Wyoming	1.04
New Mexico	0.84
Arizona*	0.05

Lower Basin—7.5 million acre-feet per year (2.4 trillion gallons)

State	Acre-feet per Year
California	4.40
Arizona	2.80
Nevada	0.30

*Arizona also gets an allocation of 0.05 million acre-feet per year from the upper basin in addition to its allocation from the lower basin.

obligation to the lower basin states stipulated by the compact, California agreed to limit its claim to the river to 4.4 MAF rather than continuing to use ever more water and thus glean ever more water rights on the basis of the prior appropriation doctrine. With this concession from California, the other states lent their political support to a bill that would authorize a great dam in Boulder Canyon and also recognize the division of water between the upper and lower basins. And thus the newly elected President Herbert Hoover signed the Boulder Canyon Project Act of 1928, authorizing the construction of what we know today as the Hoover Dam and also allocating the water among lower basin states.[16]

There was one holdout: Arizona. Ornery Arizona. The commissioner from Arizona had signed the Colorado River Compact in 1922, but the state legislature refused to do so. The other states changed the rules so that they needed only six of the seven states to ratify this compact and move on with water development, but the State of

Arizona still did not ratify the compact. Nor did the state support the Boulder Canyon Project Act, which formalized water allocation among the lower basin states, including Arizona. While Arizona had several reasons for resisting the compact for so many years, its big complaints were that the state didn't have an aqueduct and it didn't trust California.

Arizona had long battled with California over the Colorado, but Arizona always had a weaker hand because it lacked the infrastructure for putting the water to use where it was needed—near Phoenix, some two hundred miles away from the river. Soon after the compact was signed and the Hoover Dam constructed, California began sending Colorado River water off into its own hinterlands: by 1939 Colorado River water was watering Los Angeles through an aqueduct spanning more than 240 miles; and in 1942, California was irrigating the farms of the Imperial Irrigation District 80 miles from the river. As long as Arizona was unable to build the aqueduct necessary to divert its portion of the Colorado River, California could take more than its allocated share.

In 1944 Arizona threw in the towel and signed the compact—in part so it could sue California—thus formally asserting its rights to an annual allocation of 2.8 MAF of Colorado River water. But even with its water rights secured, Arizona still had no way to get the water to where it was needed, over two hundred miles away in the central part of the state. For that feat, Arizona would need an aqueduct. It would cost an enormous sum of money as well as much of Arizona's remaining hydrologic sovereignty.

What Arizona got forty years after the 1922 compact was the Central Arizona Project (CAP), at that time the single most expensive piece of legislation ever passed by Congress. The CAP eventually would take water out of the Colorado River on the state's western border at Lake Havasu (where the Los Angeles aqueduct took water in the other direction) and move that water east to Scottsdale, on to Phoenix, and then on to Tucson along its 336-mile path through the Sonoran Desert.

Arizona had been trying for decades to get an aqueduct like the CAP, but for most of those years the state had been blocked by California, which was then acting like a hydrologic and political bully. By blocking a canal for Arizona, California could continue to grow its water use and thus rely on the appropriation doctrine to sustain such use over time—or so it hoped. That tactic was finally blocked by the Supreme Court in 1963, which said that regardless of what Arizona and California did after the Boulder Canyon dam was built, the distribution between the states as outlined in the act authorizing the dam would hold: 4.4 million to California, 2.8 million to Arizona, and 300,000 to Nevada. California's water from the river was capped.

But California was not out of political maneuvers yet. With its water capped, California fought to ensure that it had the most senior priority for water in the lower basin. If things got really bad in the future, California wanted to make sure to get all of its water before Arizona got any of its. Arizona, desperate for the aqueduct, made a deal with California, its hydrologic devil. In 1968 Arizona agreed to be the junior rights-holder of the lower basin in exchange for California's political support for the CAP. All of these negotiations and compacts and new laws set the rules. The U.S. Bureau of Reclamation would do the work.

Though the Bureau of Reclamation would become the twentieth century's chief dam builder, its beginnings were humble. Started under Teddy Roosevelt in 1902, the role of Reclamation was to increase the scale of irrigation in the West beyond what was possible by local efforts alone. The bureau's early work focused on already existing irrigation projects that were marginally effective but, with federal money and Reclamation engineers, could grow to be self-sufficient.

Indeed, an enormous number of irrigation projects were well under way by the time the Bureau of Reclamation came onto the scene. While settlers on the Great Plains were notoriously self-sufficient and independent, settlers in the interior West recognized that the irrigation infrastructure needed could not be built, managed, or even justified by an independent farmer. Very early on, western settlers solved this "collective action" problem by forming groups to build irrigation systems that combined simple infrastructure, organization, and management. Some of the earliest irrigation groups were the Mormons. In 1847, within two hours of arriving at their new home in the Great Basin of Utah, these religious pioneers diverted water from the 8-foot-wide City Creek to soften the ground for plowing. Within twenty years, they had built a system of diversion dams and 277 canals that gave them enough hydrologic certainty to reliably feed their burgeoning population. They achieved this feat by developing a novel management system of irrigation districts through which property owners could locally organize and levy taxes on themselves to build and maintain canals.[17]

Additional forms of local governance and even entrepreneurial approaches to manage irrigation would soon emerge in response to the reality that the optimal size for irrigation was greater than the optimal size of a farm. By 1920 the average size of a farm in the West was just over 260 acres, of which 83 were irrigated; the average size of an irrigation organization, however, was thousands of acres. Western settlers had pioneered local governance as much as they had pioneered the use of irrigation technology, which together began to transform the West: over 3 million acres were under irrigation in the West before the Bureau of Reclamation—the federal government's extension to aid irrigation—even existed.[18]

The vast majority of these nineteenth-century irrigation systems were effective at diverting and transferring water, but their critical weakness was not being able to store water from one year to the next. The irrigation systems maximized their use of the water available in a particular year, but they were vulnerable to the vagaries of flow from year to year. If the federal government was going to put to productive use the vast lands that it had purchased and accrued since the Louisiana Purchase in 1803, then it could not leave all the work to individuals or even to local government–like irrigation districts. Rather, the federal government would have to take on the cost and effort of the more significant storage and irrigation work necessary to settle such immense expanses of arid lands.[19]

Reclamation expanded the scale of what was possible, or even imaginable, particularly by constructing the ever-larger dams that came to symbolize the agency. Large dams were the central element of water storage, the essential component of scaling up irrigation in the West. With the advent of hydroelectricity, these dams served double duty in encouraging settlement by providing both water and power. The Hoover Dam in particular was visual evidence of the scale made possible by Reclamation. The world's largest dam *and* the world's largest hydroelectric power plant at the time of construction, the Hoover Dam could store 30 MAF of water, or almost two entire years of Colorado River flow. It could then mete out that water to the lower Colorado River for use in irrigating vast farms in the desert and for filling Lake Havasu, the reservoir formed by Reclamation's Parker Dam 150 miles downstream of the Hoover. From Lake Havasu, California diverted water to the west while Arizona diverted water to the east. Reclamation provided the gargantuan scale needed to stabilize flows and deliver regular, secure water supplies to the burgeoning Southwest.[20]

In addition to the Hoover Dam, successful execution of the Colorado Compact called for a network of massive dams upstream of Lee's Ferry. A single large dam could regulate the flow of a river for a year or two; a network of mega-dams peppered along the main river and its tributaries could manage the flow of an entire basin over many years.

In the decades after the Colorado Compact, Reclamation built the Flaming Gorge Reservoir on the Green River, the Navajo Lake Reservoir on the San Juan River, the Glen Canyon Dam just upstream of Lee's Ferry, and a string of other towering dams and sprawling reservoirs on almost every other river of the upper Colorado basin. In all, the dams and reservoirs of the Colorado River Storage Project (1956) in the upper basin could hold 34 MAF of water. This sequence of dams did not just rebalance the surpluses and deficits of spring and summer flow; it rebalanced the surpluses and deficits of the region's naturally inconsistent and drought-prone climate by building up a reserve in wet years to be doled out in dry years. Such mega-dams were also essential for the upper basin states to satisfy the linchpin of the agreement, for the creators of the Colorado Compact believed the dams would ensure that the lower basin got its 75 MAF of water over any 10-year period.

By stabilizing flows and rerouting rivers into canals, Reclamation made possible a whole new hydrologic norm. No longer were rivers dangerously swollen with snowmelt in the spring, only to dry out in the summer. With mega-dams capable of storing more than an entire year's runoff and carefully sequenced networks of dams peppering the upstream river network, rivers became steady, regulated, hydraulic machines.

As canals extended like spider webs originating at Reclamation reservoirs and then crossing state boundaries, the desert Southwest became hydrologically interconnected. Simple rivers flowed downhill to the ocean. But once Reclamation had done its work, rivers of the desert Southwest no longer depended on gravity. They were pumped hundreds or thousands of feet up and out of canyons, over mountain ranges, and along hundreds of miles of canals through the desert to waiting fields and cities.

Thanks to Reclamation, states could be assured of receiving the water over which the Colorado Compact had declared them sovereign. But from there it was up to the states, the irrigation districts, and the farmers to divvy up the water. This was the essential hydrological and political shift; from taking the water out of the river to putting it to use on a field. And this was also an ideological shift: from there the question revolved around property rather than sovereignty.

CHAPTER 6

A New Water Market

Vince Vasquez spends an inordinate amount of time crisscrossing the Southwest, spending weeks on end poring over maps, digging into details of land ownership, driving tens of thousands of miles through the desert, and engaging in hours of discussions with farmers about water rights. Vasquez manages projects in Arizona and New Mexico for Water Asset Management, a private equity and hedge fund that invests in water. The firm, known as WAM, is based in New York and was founded on the idea that water should be thought of as an entirely distinct asset class. WAM invests in all things related to water, from consulting firms that have technical expertise in water and companies that manufacture pipes and valves to public and private utilities that provide water to technology companies that make sensors for measuring pollutants in water. In the drought-ridden Southwest, Vasquez is the managing director for deals in New Mexico and particularly Arizona, where WAM invests in water rights along the Colorado River—or what is increasingly being thought of as a type of "water market."

Water markets have operated in the United States since the water carriers in New York City and Chicago filled casks with water from clean sources and sold them along the city streets. A different sort of water market is possible in the western United States because of the system set up by gold rush miners in California. Because water is separable from land in the West, and is treated as property, it can be bought and sold like any other extractable aspect of land, from timber to bales of hay. In the western United States, water markets are really water rights markets: farmers with water rights can be paid to watch water go by their diversions and forgo their right to divert that water for their own fields, either short term—as a water lease— or permanently, as a sale. These farmers can and occasionally do sell or lease their water rights to other farmers; or, more lucratively, they sell or lease to cities or industries that often are willing to pay much more for water.

Water use in the West has always been dominated by agriculture. Even though most of the population growth in the western United States—which has more than quadrupled in the past century from 70 million in 1900 to over 320 million in 2014—has primarily been in cities, irrigation still makes up more than 80 percent of water use. This is due to the appropriation doctrine: because agriculture arrived first, the most senior water rights are invariably held by farmers, who often sold or passed them down to their children along with the land. The result is that farmers own most of the hydrologic property in the West.[1]

As urban population grew and municipal demand for water increased, cities have had to get in line for water with everyone else, behind a long list of senior agricultural water users. But cities and big corporations in industries such as mining are typically willing to pay a premium to cut in line. The low supply and high demand sparked the development of water rights markets as early as around 1890, when the City of Colorado Springs went to court over one of the state's earliest sales of water from irrigator to municipality. Although it had a water right and pipeline intake on a tributary to Fountain Creek, Colorado Springs needed more water for planned growth and purchased additional water rights from an upstream irrigator. When this sale of water rights was challenged in court because it would involve changing the use of water from irrigation to municipal purposes, the Colorado Supreme Court said the sale was legal because a water right is "the subject of property and may be transferred accordingly." With this case in Colorado, and similar treatment of water in cases in other western states, water rights officially, legally became a tradeable commodity.[2]

We buy and sell bread by the loaf; we buy and sell oil by the barrel. Water is bought and sold by the acre-foot, and at the turn of the twentieth century the market for water began growing rapidly. By 2015 about 1.8 million acre-feet (MAF) of water was traded each year, and the total value of those trades reached just under $800 million. While most of the market activity was in California, Idaho and Arizona were hubs of water swapping as well; not to be outdone in anything, the water market in Texas was growing rapidly. Most of the sales in the West are from agriculture to municipalities, at prices anywhere from under a hundred dollars up to a few thousand dollars per acre-foot of water. But specific transactions can reveal just how valuable water can be to specific buyers under specific conditions. In 2014 one of the world's largest mining companies, Freeport-McMoRan, spent $1.3 million for 90 acre-feet of water—a shade under $14,000 per acre-foot—for operations at a copper mine in eastern Arizona. This would be a prohibitively exorbitant price for agriculture, but it is just a necessary cost of doing business for a profitable international mining operation dependent on water.[3]

Vince Vasquez's work for WAM is finding opportunities to buy water rights—and often the associated land—with an eye toward selling them at a profit sometime in the future. One of WAM's earliest ventures into the western water world was in Arizona in the late 1990s, before Vasquez joined the firm. The founder of WAM had his eye on water from the recently completed Central Arizona Project (CAP), which had just started delivering water to the Harquahala Valley Irrigation District. In Harquahala, the land itself was not particularly valuable for many crops. At the time, cotton was selling at 39 cents per pound, but it cost 42 cents to grow with irrigation.

However, Harquahala sat just outside the western fringe of Phoenix's suburban sprawl, where water was in high demand. Land that initially sold for $400–600 per acre went up to $1,000 per acre once the CAP water arrived. When WAM bought the land and associated water rights, it used some of the water to irrigate agriculture, but sold much of it for more valuable—or at least more profitable—use in cities and industries.

Vasquez is looking for similar deals, but with an eye toward the value of the farm as well as the water. Vasquez's job is to find farms at a reasonable price, improve those farms with updated infrastructure, and then either continue farming with increased productivity or separate the water from the farm for sale. A primary driver of the land value, and water, is the seniority of water rights. But farmers with senior water rights are not all looking to sell their land or their water. Just like any other investor looking for willing sellers, Vasquez has to know where to look—and, as important, when to look.

The perfect deal for Vasquez is to find a farmer looking to sell out completely. This situation gives Vasquez maximum flexibility for both the land use and the water use; he could sell the water outright, or he could update the farm to allow for growing higher-value crops with the available water. More often, farmers want to keep farming but need an infusion of cash for maintenance or new equipment; or they need a nest egg for retirement while continuing to farm for a few more years. Farmers may not be able to switch to higher-value crops simply because it takes a lot of up-front cash that they may not have. Or, as Vasquez captures it while driving through the western Arizona desert, "Farmers are water rich, but cash poor, whereas we have a checkbook and an appetite for risk." For farmers who want to keep the land but need cash, Vasquez structures deals to buy the land but then lease it back for a period of time to allow the previous owner to continue farming.

This business model does not translate into the deal known as "buy and dry"—buying the land, selling the water, and abandoning the farming operation in the process. Rather, Vasquez buys the land

and makes the farms operational. He spends his days doing the nitty-gritty work of updating farms that he owns. On a typical day, he negotiates a deal for clearing brush from a few hundred newly acquired acres, then drives a mile away to plan a series of new irrigation works on another property, then heads into town to buy a power pole to get power across a rural backroad for another property. All the while, at an adjacent property Vasquez recently bought, enormous tractors are leveling several hundred acres to laser-precise flatness in preparation for a series of new irrigation infrastructures there. This day-to-day work is done to make the farms that WAM has bought fully operational and as hydrologically efficient as possible.

All of this work raises a question: Why pour so many resources into making farming in the Arizona desert more efficient if you're in the business of selling water? First, these particular water rights are connected, by the Colorado River and the Central Arizona Project, to the ever-sprawling suburbia of Phoenix. This means that every acre-foot of water Vasquez saves on the farm is transferrable—sellable— in Phoenix, hundreds of miles away. But perhaps more important, the farm Vasquez is updating is along the Colorado River, and it has been irrigated with Colorado River water since before 1922; before the Colorado Compact. The water rights from this farm are referred to as "present, perfect rights"—the most senior, inalienable water rights of the Colorado River system. Because Arizona had to move to the back of the water line behind California to get CAP funded, the water flowing through CAP is always in danger of being cut off during a drought. But the water rights that Vasquez is developing, and potentially selling in the future, are equal to some of California's most senior rights and thus are inordinately valuable in Arizona.

Why all the time, money, and effort on clearing brush and building canals when the intention is, presumably, to sell the water anyway? Because Vasquez needs to demonstrate that he is putting the water to beneficial use and thus secure his rights before he can sell them. And the amount of work he puts into upgrading the water infrastructure and basic farming equipment removes all doubt that he's putting the water to use. Additionally, water buyers like Vasquez view highly reliable water rights as undervalued to date and are betting that the longer they hold the rights, the higher their value will become. Because the risk in these water deals is quite large, buyers may not be able to sell the water or may not be offered a good price. Vasquez says, "We may have a farm for a very long time." But WAM and other water buyers have capital that other farmers do not. Even without selling the water, they can get the farming operation up and running and providing high economic returns to offset the cost of the purchase on its own. And what's more, Vasquez can constantly pursue other water deals in the area while running the farm. Rather than selling all the water rights, he just lets a portion of the farm lie fallow. This plan generates water savings that Vasquez can then lease or sell.

From a bluff overlooking the Colorado River valley, Vasquez's renovated farms are bright, emerald-green square patches amid the dusty wasteland of the surrounding desert. All the fields are planted with alfalfa hay. Vasquez's farms grow hundreds and hundreds of acres of alfalfa, but they account for only a small portion of the thousands and thousands of acres of alfalfa grown here, in one of the most arid regions on earth. Each acre of alfalfa requires 5 acre-feet of water, the equivalent of 60 inches of rain—in a region that typically gets only 5 inches per year. Aside from cotton, which remains marginally profitable, alfalfa is the primary crop in this region. Its production is steadily increasing to meet a seemingly inexhaustible demand.

In 1993 California passed Wisconsin as the leading milk-producing state; in 2014 California produced 5 billion gallons of milk, almost doubling the amount produced in "America's Dairyland," Wisconsin. Today California alone produces over a fifth of the nation's milk. And the alfalfa being grown in arid Arizona enables California to be a dairy powerhouse. Alfalfa grown in the spring to be a farm's first cutting is high in protein—ideal for dairies. The second cutting, later in the summer, is less valuable to dairies but can be exported internationally to feed livestock. The proximity of Vasquez's farm, and the others around him, to international ports via interstate highways allows them to easily export their summer crops to the global markets. Container ships coming from China with manufactured goods don't have to return empty; instead, they return loaded with hay. Farmers irrigate hay in the Arizona desert at great hydrologic expense so that milk can be cheap in Los Angeles or beef can be affordable in Beijing.

The influence of this globalized alfalfa market is evidenced by the modern desert landscape around a river, from the lower Colorado River valley of Arizona to the Yakima River valley in central Washington. Almost any western river valley is a patchwork of irrigated alfalfa hayfields punctuated by building-sized towers of hay bales averaging over three stories tall, 20 feet wide, and almost 200 feet long. Cryptic numbers and arrows painted on the sides of the towers of hay indicate which hay has been contracted for pickup and will be shipped somewhere across the Pacific Ocean. The view from a bluff above the Colorado River reveals towers of hay, scattered across the landscape up to the horizon, resembling tall grain elevators in the Midwest.

There's another reason alfalfa is a popular crop here in Arizona and in much of the West: its particularities make it nearly perfect for water deals. Alfalfa production is optimal when planted on a 3-year rotation—in other words, it does best if the field lies fallow one out of every three years. For Vince Vasquez, this means that in any given year, a third of the WAM properties growing alfalfa necessarily are out of production. This makes at least 5 acre-feet of water per acre of fallow land available for lease to the highest bidder. On a square mile—640 acres—of land, fallowing one-third would free up just under 350 million gallons of water per year—enough water to supply about 10,000 suburbanites in Los Angeles or Phoenix. Alfalfa yields profits for western farmers, and it gives them enormous hydrologic flexibility from year to year. This is something that fixed crops like almond and apple trees or vineyards, all of which require irrigation every year, don't provide.

Water markets often come under fire from opponents because of their potential for deals leading to buy-and-dry situations, when firms sell all the water and abandon the land to a dusty, useless future. But the approaches that WAM and Vasquez are using—keeping farms operational and practicing rotational fallowing to make some water available for sale or lease—can generate multiple benefits. Thanks to these more innovative approaches, cities can meet at least some of their growing water demand, and farm communities can stay in business and enjoy continued productivity.

In other regions, like the trout streams of western Colorado and the salmon-rich streams of the Columbia River basin, similar approaches have been used to release water from agriculture and keep it in the streams for environmental purposes. Nonprofit environmental groups increasingly use water markets as a way to accomplish their conservation goals. This is an underappreciated aspect of treating water as property. Despite criticism, the water market has proved to be a mechanism for realizing water conservation benefits—as long as environmental groups are willing and able to pay as much for water as other farmers, cities, or water traders like Vasquez. Markets allow water to be reallocated to where it is most valued, and some basins yield enough water to be bought and sold to meet the needs of different interests. However, it is becoming apparent that the system of water distribution used for the Colorado River, and the systems used for other rivers, are not nearly as viable as initially thought. The result is shortages that were not foreseen almost a century ago.

Dendrochronologist Erika Wise, at the University of North Carolina, has an unusual lab: a very high-end carpentry shop filled with power sanders, band saws, stumps, and sawdust. Her latest piece of equipment is a new sander. Her new graduate student was a catch—partly because he's wickedly smart and partly because his dad is a woodworker, and this student knows his way around wood. This skill is essential—wood is what dendrochronologists spend their days studying.

Many of us have counted the rings of a tree stump to figure out how old the tree is. Dendrochronology—the science of tree rings—takes this simple method many, many steps further. Trees grow rings differently in response to different conditions. Diseases and forest fires leave telltale marks. Dendrochronologists can observe variations in tree rings throughout a region and estimate when these types of events occurred, how severe they were, and how widely the fire or disease spread. Tree rings also reflect water availability, which is what interests Erika Wise.

Wise's office is littered with slices of old trees that have accumulated through the years. One—from a limber pine—she saved because it was her first time using a chain saw. Other slices—cross sections—from trees are scattered about her office. Each of these slices has been sanded down to better show the hundreds and hundreds of densely concentric rings. A few buildings away from her main office is her lab, tucked away in the basement of the botany building. There she carries on the business of analyzing tree rings, which is part adventure and part mind-numbing repetition: "The hardest part is finding new sites. If a tree is easy to see or easy to get to, it would already be cut down. For dendrochronology, you need old trees, and old trees, they are usually not in . . . well . . . safe places." By site, she means a collection of old trees, and by old trees she means trees that are five, six, even seven hundred years old. Old trees tend to be far away from roads, on really steep slopes or cliffs. Old trees aren't necessarily the biggest trees either, which means that Wise has to sample a lot of trees when she thinks she's found an old patch to make sure she gets a sample from the really old trees that might be in the patch.

Sampling means coring, the process of slowly turning and boring a large hollow drill bit into a very old tree and then carefully pulling the bit out. After removing the bit from the tree, Wise removes the resulting cylindrical core from the tree. Each core is several feet long—about the radius of the tree—and about half an inch in diameter. Wise can obtain these cross sections to analyze without felling or otherwise seriously damaging the tree. She spends months each summer hiking the backwoods of the West in search of old trees and taking days to core them. During her last sampling run, she sampled about 40 trees per site, taking at least two cores per tree. "I often overdo it," she says, "I mean, it's a lot of work to get there, and you never know which tree is going to be the old one." During her last summer, she had eight sites to sample—enough to send hundreds and hundreds of cores back to the lab.

Back home, she is in the midst of renovating the lab while analyzing the cores. Carpenters and painters on one side of the lab are sanding the new wood of a lab bench; meanwhile, on the other side, Wise sands the centuries-old wood of a tree core. The sanding is crucial because some of the rings in a tree may be only a couple cells thick. And those rings are potentially the most interesting because they are the drought years. So Wise and her students sand each tree core six or seven times, using increasingly fine-grained paper, until the last sanding with jeweler's paper. From there the mind-numbing work begins: they place the core under a microscope and measure the width of each tree ring to the nearest 0.001 mm; that is, to the nearest thousandth of a millimeter. So for a single site that includes forty trees of about 500 years old, with two cores per tree, Wise and her students measure tens of thousands of tree rings. It's data painstakingly won, earned by bending over a microscope while counting and measuring innumerable tree rings in a basement lab. As she acknowledges with a laugh, "It's really hard to keep hourly undergraduates for very long working in the lab." But the results can be striking.

Wise did some of her earlier dendrochronology work on the upper Snake River in Wyoming and Idaho. She collected some cores herself and used some collected by others to build a chronology of tree rings in the West that covers a period of just over four hundred years, from 1600 to 2005. She then used some streamflow data that had been collected on the Snake River since 1910 and overlapped the tree-ring data from 1910 to 2005 with the streamflow records from the same time. Through this work, Wise established the relationship between tree-ring widths and annual streamflow. She found that when streamflow is large, tree widths are large, and vice versa. But her meticulous efforts in the lab counting and measuring rings allow Wise to then work out the relationship between ring width and streamflow with a high degree of statistical precision. Knowing this relationship, she can use just the tree-ring data to calculate streamflow.

Wise's system allowed her to compare the droughts of the twentieth and twenty-first centuries with those of the past four centuries and then ask some rather provocative questions. For instance, how exceptional was the Dust Bowl drought of the 1930s in the context of the other four hundred years? Based on Wise's reconstruction, the 1930s drought was unusual; but several other droughts were comparable, and a couple were even worse. At the time she wrote the paper, the 2000–2005 drought was on everyone's mind because of how long it had lasted; Wise used tree rings to show that two other droughts— one in the eighteenth century and one in the seventeenth—had been even longer. For water managers in the Pacific Northwest, tree rings began to shed new light on just how severe things might actually get.[4]

Wise did her PhD work at the hub of dendrochronology—the University of Arizona. Scientists there have been compiling and studying tree rings for decades to better understand past climates of the desert Southwest, particularly around the Colorado River. One study got people especially concerned. In 1976, Charles Stockton and Gordon Jacoby published a fairly obscure report in the *Lake Powell Research Project Bulletin*. Like Wise's study, it correlated the flow of a river to tree-ring data.

Stockton and Jacoby had measured a series of cores from trees located throughout the Colorado River basin and correlated their ring measurements with the longest record available of river flows: those at—where else—Lee's Ferry. Like Wise, they were able to look at drought frequencies and durations; but their findings on droughts were not nearly as profound or startling as their discoveries about floods. They found that "the early part of the twentieth century (1906 through 1930) was one of anomalously persistent high runoff from the Colorado River Basin, and that it apparently was the greatest and longest high-flow period within the last 450 years." In nonscientific terms, Stockton and Jacoby were saying that the decades of data used to estimate flow on the Colorado River and then divvy it up in the Colorado Compact were in fact the wettest years in half a millennium. The basis of the Colorado Compact was an estimated mean annual flow of somewhere around 16.5 MAF per year; tree-ring data over a much longer period of time suggested that a more realistic estimate would be about 3 MAF per year less than that.

After Stockton and Jacoby dropped this bomb on the western United States, later studies using tree rings revealed even worse news. Not only was the early twentieth century consistently coming up as being a period of "exceptional wetness," but as other dendrochronologists expanded to similar work throughout the West, they began to show that the droughts of the Dust Bowl and the drought of 2000–2005 were not as rare on the Colorado as people had assumed: "Overall, these analyses demonstrate that severe, sustained droughts are a defining feature of the Colorado River."[5]

The results of these tree-ring studies have profound implications for the states upstream of Lee's Ferry. Because the Colorado Compact commits the upper basin to delivering 7.5 MAF per year to the lower basin, the inevitable shortfall of flows will come out of the upper basin's allocation, forcing the entire region to rethink its hydrologic future. The potential impact of this collision between the realities of dendrochronology and the Colorado Compact was best captured by Richard Lamm, governor of Colorado during the 1977 drought, who said, "The most disturbing day I spent in my 12 years as governor was having tree rings explained to me." Arizona faces a similarly problematic situation because, aside from a few hundred acre-feet here and there of present perfected rights, the state gave California seniority in exchange for funding the CAP. In a sustained drought on the Colorado, when Lake Mead falls ever lower and Lake Havasu dwindles, California will continue to divert its 4.4 MAF, and Arizona will have to make do with the leftovers.[6]

All of the historical climatological changes the dendrochronologists have documented were natural oscillations driven by ocean currents and atmospheric circulations, cycles that have gone on for hundreds or even thousands of years. Human-induced climate change is likely throwing this system out of whack, or at least amplifying the natural oscillations. Wise and other dendrochronologists and climate scientists predict that future droughts will be drier, and future floods will be wetter. Infrastructure built for past climates may no longer be adequate to store the floods of the future or to supply water through the length of droughts in the future. Even slightly warmer temperatures in the West cause more precipitation to fall as rain instead of snow and cause the snowpack to melt sooner. All of this means that planners and engineers cannot rely on snow to store water in the spring or snowmelt to provide water in the summer, and more reservoirs or larger reservoirs will be needed to store the same amount of water. All the while, the population continues to grow.[7]

Stretching along the lower Colorado River, south of where Vince Vasquez was buying farms to sell water, are the irrigated tribal lands of Native Americans. The fields here differ from those to the north in that they're larger and less fragmented by any fallow fields or even buildings. They are also planted with lots of cotton, which right now is less profitable than alfalfa.

It is unusual in the United States for tribes to seemingly have the upper hand in natural resources. The tribes have been pushed around and taken advantage of since the first European explorers "discovered" America. Even today, on sovereign reservations, tribes are rarely left enough to scratch out a living. But in the case of Colorado River water rights, the tribes finally have some advantage. On the lower Colorado, the Supreme Court said that the tribes were entitled not just to reservations of land but also to water: "enough water to irrigate all the practicably irrigable acreage on the reservation." But importantly, this was not going to be new water for the states where tribes were located: the water for reservations comes from each state's allotment; for example, tribes in Arizona would get water from Arizona's share. The Court also said that the priority date of a tribal water right dated to when the reservation was created, thus giving many reservations some of the most senior claims on the river. Tribes had regained some of their water sovereignty.[8]

On other western rivers, including the Klamath, tribes have been similarly exerting control over water rights—both the right to use water for irrigation and the right to have it left in the river to sustain fish and river-dependent wildlife. As far back as a 1908 Supreme Court case, it was recognized that tribes have "reserved" water rights. These rights belong to the tribes by virtue of the reservations, so they are similar to eastern riparian water rights: they didn't have to be used for irrigation to be secure; they went along with the land regardless. But it was never clear whether these water rights were enforceable if the tribes were to make a claim.

The shifts in climate, growth in population, and never-ending demand for water have increasingly tested these rights throughout the West. Increasing water stresses also put to the test all the previous assumptions in the Klamath River basin about water rights, property, and sovereignty. Farmers of the Klamath basin, like Tom Mallams, had assumed that water was their property. The state had granted them property rights to water, with associated dates of priority, and the farmers had built their farms and their lives around the assumptions built into these water rights. The Oregon farmers' approach was no different from that of farmers in Kansas building a farmhouse and barn based on the amount of land—property—they own.

But when the extreme droughts of the twenty-first century hit, it became painfully clear that water rights were not as secure as farmers in Oregon had thought—because the state of Oregon was not as sovereign over water as they had thought. In 2013 the tribes on the Klamath decided to assert their legal rights to water by "making a call" for water—requiring farmers like Mallams to cut back their upstream diversions significantly and instead let streamflow go downstream unused to sustain the fish central to tribal culture. Mallams was frustrated partly because his state-given rights were being superseded by tribes, whom he associated with the federal government.

In his calm, measured tone, Mallams walked through his reasoning: "That water belongs to the State of Oregon. Water is the property of Oregon. But now the right to make a call on water has been handed to the federal government. When the tribes make a call on the water, they make it through the Bureau of Indian Affairs, and that is the federal government. The federal government now controls the water. The states have capitulated to the federal government." For Tom Mallams the question is, if the State of Oregon must relinquish the power to decide who owns water in times of extreme drought, then what sovereignty does the State of Oregon really have?

Conversely, tribes like Leaf Hillman's, which had assumed the worst after centuries of broken promises, were given the most valuable asset of the West: senior water rights. Through myriad lower court and Supreme Court decisions, and increasing societal recognition of the legitimacy of their claims to the use of water, tribes had become the hydrologic sovereigns of the West. Though their reservations were still small compared to their original areas, their situation was beginning to improve. Hillman says that if the Karuk Tribe is a sovereign nation able to set its own rules, except for water, then is it really sovereign? As he captured it, "There can't be gradations of sovereignty. If there can be limits put on sovereignty, then it isn't sovereignty."

Klamath farmers were living in a new world, and they knew it. As we shook hands before I left his office, Mallams gazed out the window at what seemed to be an ever-cloudless sky. He enjoys being a commissioner but knows that eventually he will need to retire and give up some of his role in fighting all these water wars. He wants to go back to farming, "But I might not have the water to go back. The way things are now, there is no security for water. Zero. You have no security."

This is the reality of water in the West: a zero-sum game. The system of sovereignty and property has been used for centuries to establish who controls what—or, in the West, who controls water. Droughts, disappearing species, and burgeoning population all increase demands on an already overextended system. Yet the changes in sovereignty and property have an equal or even greater effect on the farmers, tribes, and water brokers of the West. And they raise questions about the future of western rivers.

PART THREE

TAXATION

CHAPTER 7

Running Water

Being able to blithely drink water from just about any faucet in the United States without concern is one of the greatest achievements of American society. While we may not always like the taste, we do not typically have to think twice about drinking the water from whatever water fountain is available at a rest area, an airport, or a bus station. We don't even have to worry about whether the water is potable, which reflects an accumulation of technical sophistication that has enabled an infrastructure system that sprawls out into the landscape to gather water from diffuse sources, draw it through pipes and pumps and canals, purify it, and then distribute it through more pipes and pumps to every home, office, and city park water fountain. When it comes to wastewater, we casually flush toilets expecting that pipes and sewers will do whatever it is they do, underground and unseen.

A functioning water system—clean tap water and a reliable sewer system—is a staggering accomplishment of technology. But water systems are equally the result of the development and use of finance by governments—to develop a budget, borrow money to finance large investments, and then secure the revenue to pay their debt. Running a government, whether a federal agency or a city sewer, means running a budget. And budgets reflect power; taxing, revenue, debt, and spending all indicate society's acceptance of the government exerting some monetary control. Budgets are simply mechanisms for making economic and political choices.

Since the founding of the United States, political ideology has been inseparable from fiscal ideology. A central disagreement between Alexander Hamilton and Thomas Jefferson was what to do about the debt accrued by the individual states in fighting the Revolutionary War. Jefferson wanted to leave the states to fend for themselves financially while Hamilton wanted the federal government take over the debt from the states. This disagreement in proposed fiscal policy reflected a larger ideological divide. As both Jefferson and Hamilton recognized very early on, government debt is an indicator of government activity because taking on debt requires shouldering the obligation to collect sufficient taxes—revenue—to pay off the debt. Whereas Hamilton's approach would require a federal tax to pay off the debt it took from the states, effectively increasing the federal government's role, Jefferson believed in limiting the role of the federal government as much as possible. In 1811 Jefferson praised the tariff (a tax on international imports) alone as the sole legitimate source of federal government revenue. By taxing the rich and leaving the small farmer untouched, the tariff would be an effective, redistributive fiscal approach, particularly if the revenue was used for projects like canals and roads that benefited farmers:

> We are all the more reconciled to the tax on importation [tariff], because it falls exclusively on the rich, and with the equal partitions of interstate's estates, constitutes the best agrarian law. In fact, the poor man in this country who uses nothing but what is made within his own farm or family, or within the United States, pays not a farthing of tax to the general [i.e., federal] government, but on his salt; and should we go into that manufacture as we ought to do, he will not pay one cent. Our revenues once liber-

ated by the discharge of the public debt, and its surplus applied to canals, roads, schools, etc., the farmer will see his government supported, his children educated, and the face of his country made a paradise by the contributions of the rich alone, without his being called on to spare a cent from his earnings.[1]

Per Jefferson's agrarian and state-centered ideology, limiting federal debt would beget limited federal taxing and spending and thus limit federal activity and power altogether. For him, decentralized finances were a tool to ensure decentralized power. Indeed, as president, Jefferson developed a religious commitment to a balanced federal government budget with no federal debt—the fiscal extension of his limited government ideology.

This tension between Jefferson and Hamilton over budgets, and which level of government should be responsible for debt and taxes, was the ideological and fiscal starting point for the United States. The whole economic history of the United States is the saga of negotiating the fiscal roles and responsibilities of the different levels of government in providing the most basic of services for their citizens—the water supply and sewer systems.

When the 38-year-old Ellis Sylvester Chesbrough set out from Boston and headed west toward Chicago in 1851, he was crossing a landscape in the throes of dramatic changes both fiscal and physical—much of which he had witnessed firsthand. Chesbrough had grown up in Baltimore watching his father work for the Baltimore and Ohio Railroad Company, which was responsible for the Mid-Atlantic region's most ambitious building project of the early nineteenth century. Ellis Chesbrough followed his father's lead, and at age 15 he began working as a surveyor around Baltimore. In a few short years, he rose through the ranks to serve as an assistant engineer. Like many of the civilian engineers in his time, Chesbrough was self-taught, but he was mentored by the more experienced men who did the actual engineering work on enormous engineering projects ranging from railroads to canals.[2]

This was an ideal era for a young, self-made engineer like Chesbrough, and the projects he worked on over the years show where and when the economy was growing—or shrinking. When Chesbrough began his career in the early 1830s, the federal government was largely inert and its engineering work was limited to a few harbor projects. In contrast, large engineering and building projects funded by the states were sprouting up all across the United States. For engineers like Chesbrough, working for a state government was the way to gain experience as an engineer because the different states were frantically building canals, railroads, and turnpikes in efforts to outdo each other.

The projects being undertaken by the states were certainly necessary, but they were also driven in part by the fiscal model adopted by states in the early nineteenth century: asset income. Asset income entailed financing projects that could pay for themselves from revenue generated by the direct use of those projects. State governments sold bonds to generate funds for constructing infrastructure projects. People using the infrastructure would pay tolls or fees, which would then be used to pay off the state debt—that is, the bonds. As these public works projects became sources of revenue, the states would be paying back bondholders and simultaneously growing the state economy.

The best example of this type of investment was the Erie Canal, a model that all other states looked to in justifying their frantic building campaigns and associated debt accrual. The Erie Canal had been remarkably successful in achieving the ideal of a self-financing building project that grew the economy while it grew itself: it increased the market for commodities, which drew more settlers to the region, whose use of the canal increased its revenue, which allowed the state to repay the bondholders on schedule and in full. The returns on Erie Canal bonds were so secure that the bonds themselves were used as currency, sold throughout the United States and even in Europe. The allure of the asset income approach for states was that it reduced the need for direct taxes such as property, poll, or income. State investments in public works then could both increase economic productivity and greatly reduce general taxes by targeting only those using the specific state-funded assets.[3]

Knowing that New York's coffers were burgeoning, all other states enviously went into the business of internal improvements. But the necessary initial costs of designing and building infrastructure meant that by selling bonds, states also went into the business of taking on debt. Indeed, just as Jefferson had envisioned, the states rather than the federal or local governments were the hub of financial activity. By the 1830s the state share of government debt was 86 percent of total government debt, and federal debt was only 1.5 percent. Local debt accounted for the remainder—less than 13 percent. And, just as Jefferson had hoped, the federal government derived its revenue primarily from import taxes on foreign manufactured goods—tariffs to protect American industries.

Relying on asset incomes was a high-risk fiscal strategy for states. Many states incurred considerable debt to finance the canals, which in turn became primary sources of state revenues. After an initial love affair with all proposed canal projects, investors began to realize how exceptional the Erie Canal was; few other canals in the United States were used as heavily as the Erie. This lack of use in other canals led to declines in revenue and challenged states to pay the debt they had incurred, largely in constructing the canals: nationally, the total value of state bonds issued in 1820–1824 was $13 million; but that figure swelled to over $108 million in 1835–1837, and over half of it was for canal building alone. With so much state debt and essentially no regulations on finance, the first great financial panic hit the United States in 1837 during that era's technological boom— the first tech bubble.[4]

Much of the credit taken on by state treasuries to pay for canals was extended by foreign investors, most often in London. As canal and river companies began to fail, leaving canals incomplete or in disrepair and thus reducing their use and revenue generation, states defaulted on their debt payments: Maryland, Illinois, Indiana, and Pennsylvania. Virginia took on enormous debts from foreign investors throughout the early nineteenth century to finance river and canal companies, and the state did not repay its debts in full until 1966.

The British developed a jaded view of Americans' ability or willingness to pay their debts. The *London Times* denounced all Americans for the defaulting by some states, and by 1842 the U.S. government was not able to float a bond in European money markets. What's more, the United States had become the butt of English rhymes, which jeered at the American habits of both repudiation (as default was called) and slavery:[5]

> *Yankee Doodle borrows cash,*
> *Yankee Doodle spends it,*
> *And then he snaps his fingers at*
> *The jolly flat who lends it.*
> *Ask him when he means to pay,*
> *He shows no hesitation,*
> *But says he'll take the shortest way*
> *And that's repudiation . . .*

Chesbrough was on the front line of the financial collapse, largely because he was an engineer. Just as software engineers were unemployed after the 2001 tech bubble collapsed, public works engineers were unemployed after the 1837 state debt collapsed. Chesbrough, now in his early thirties, moved back home with his father.

In the mid-nineteenth century, opportunities on large infrastructure projects that spanned states were slim to none for engineers like Chesbrough. Instead, demand for engineering projects began to come primarily from the nation's rapidly growing cities. At that time, Midwestern cities in particular emerged as links between the Atlantic and the Great Plains, the Great Lakes, or the Ohio Valley. The growing network of railroads funneled a never-ending stream of grain, beef, iron, coal, and timber, all of which were processed and handled in the rapidly growing metropolises of Chicago, Cleveland, Detroit, and Milwaukee.

While the mid-nineteenth century is often envisioned as the era of California prospectors and frontier settlers out on the prairies, these were the exceptions. The real population growth at the time was in the overcrowded streets and alleyways of cities. The U.S. population grew from 23 million in 1850 to 106 million in 1920—a growth of over 350 percent compared to the world's population growth of 55 percent. The typical American city doubled in population every twenty years over the last decades of the nineteenth century, and in the Midwest that growth was further accelerated: Chicago's population doubled approximately every decade from 1860 to 1890, while Cleveland's population almost quadrupled from 1860 to 1880 and then more than doubled again between 1880 and 1900.[6]

Along with this spectacular demographic change came an equally enormous but far less appreciated fiscal change. The financial panic of 1837 brought about the first fundamental fiscal reorganization in the United States. After the collapse of canals and turnpikes, state bonds were no longer viewed as secure investments. Potential transportation customers abandoned the often unfinished canals for the far more convenient railroads, eliminating revenue for the canals and driving them out of business but also undermining the fiscal model of asset income as a basis for government budgets. In response to the panic, many states revised their own constitutions to cap the amount of debt they allowed themselves to take on. In the place of states, cities—municipal governments—grew in financial importance as investors increasingly turned to municipal bonds, essentially investing in the future of cities. And no city seemed to have a brighter future than Chicago.

The rapid growth of Chicago was due largely to its glacially carved geography, which proved both a blessing and a curse. Notably, Chicago sits at a remarkably short and flat distance—ideal for portage—

between the Great Lakes and the Mississippi River, providing trappers, farmers, and settlers with a quick link between the nation's interior and the Atlantic economy via the Great Lakes or the Erie Canal. Chicago's front porch is Lake Michigan, and its back door is the Illinois River, which drains into the Mississippi. At the end of the last ice age, Lake Michigan drained directly into a then enlarged Illinois River and on into the Mississippi. But as the glaciers receded, they left behind a hint of a ridge between Lake Michigan and the Illinois River. Without the ridge, Lake Michigan would have continued to drain directly into the Mississippi basin. But with the ridge—a minor topographic anomaly—the two great waterways were just barely separated.[7]

In 1855, when Chicago went searching for "the most competent engineer" available to grapple with this geography, Chesbrough rose to the top of the list. The situation he found in Chicago was appalling. Like many other American cities at the time, Chicago was plagued by its highly decentralized approach to urban sewage: privy vaults. Essentially, privy vaults were urban outhouses built anywhere space for a hole in the ground could be found amid the increasing density of the urban landscape. Some of them were built in basements, some behind buildings, and some so that they drained directly into streams or rivers. Everyone was left to their own devices for the disposal of human waste. The only role of the city government in this system of waste management was to arrange for these vaults to be cleaned out periodically, using dippers, buckets, casks, or pumps to scoop out the human waste. The waste—or "night soil"—would then be carted to the outskirts of cities and either dumped into a nearby stream or sold to farmers who used it as fertilizer that eventually washed downslope into creeks and streams.[8]

The privy vault system was a functional response to pollution but not appropriate for the scale of rapidly industrializing cities in the late nineteenth century. Cities across the United States were concentrating people, and the factories that employed most of the city residents concentrated them even more dramatically during the workday.

A typical New England mill village of 1,500 to 2,000 people generated about 500 gallons of urine and half a ton of fecal matter. When people were accumulated into industrializing landscapes, the numbers started to seem nonsensical: the small industrial towns of Holyoke, Chicopee, Springfield, Hartford, and New Britain contributed over 42 tons of fecal matter and almost 46,000 gallons of urine to the Connecticut River daily.[9]

Cities located on rivers had an immense advantage—at least their sewage was naturally whisked downriver. Chicago produced much more waste than the small industrial towns, but then that waste didn't drain. The city's flatness, which had been so attractive to early settlers, made the simple act of getting water out of the city an enormous task. Without the aid of gravity, sluggish water backs up and seeps into nooks and crannies, muddies soil, weakens structures, and grows bacteria; getting water out of a city is as important to the growth of great cities as getting water in. Drains and ditches were part of the earliest work done in Chicago; but without much natural slope to move the water away, they proved ineffective. As Chicago's growing population dumped its waste into the open drains, it backed up into the putrid streets, drains, and basements.

Chesbrough's comprehensive approach to Chicago's troubles set it apart from all other plans the city had tried. He intended not just to construct new drainage and sewer lines, but to apply a systematic approach to control where and how quickly the water drained. After surveying the situation, Chesbrough proposed four options to the city: (1) drain the city into the Chicago River, which would then flow into Lake Michigan; (2) drain the different streets and gutters directly into Lake Michigan; (3) drain the city into artificial reservoirs to be pumped out and used as fertilizer on nearby farms; or, most dramatically, (4) drain the city into the Chicago River and then reroute the Chicago River backward, away from Lake Michigan and toward the Illinois River to the south.[10]

Chesbrough's options balanced all the elements of what would be the key qualities of the city engineer: creativity, specificity, and sensitivity to fiscal concerns. All of the options were staggering in their scope; but Chesbrough believed the last, costliest proposal—rerouting the river—was the only long-term option viable for the city. Despite this, due to the projected expense—and because he thought that option would be necessary only if the city ever reached the then unthinkable five times its current population—he advised against it at the time.

The city took the simplest option: drain everything into the Chicago River and on toward Lake Michigan. Yet even this simple approach would require a greater slope than currently existed. This was a constraint not of infrastructure, but of topography. Chesbrough set out to rebuild the city from the sewers up, as well as rethink the way sewers were used.

There were, and are, two types of sewers: storm sewers and sanitary sewers. Storm sewers drain surface water from rainstorms; street gutters, for instance, collect the enormous quantities of water generated during rainstorms and then route it into ditches and pipes toward the nearest stream, river, or lake. Sanitary sewers collect and transport human waste, including that from business and industry. In nineteenth-century Chicago, storm sewers were small, rudimentary, triangular wooden troughs that ran along the roads and fed into the sewer mains—composed of hollowed-out logs. Sanitary sewers were rare in Chicago at that time; when they existed at all, they were merely pipes and canals tasked with carrying away the voluminous human waste. Instead of sanitary sewers, most of Chicago relied on thousands and thousands of privy vaults, which Chesbrough soon came to despise.

The first novelty of Chesbrough's vision for Chicago was to integrate the two types of sewer systems, creating one "water-carriage system" that would collect sewage from buildings and streets and use the water flow from the storm sewers to carry away the sanitary sewer waste. Combined sewer lines that carried water from storms in the same pipes that carried sanitary wastewater would later become the norm, following Chesbrough's example in Chicago.[11]

Beyond this central innovation, the project began to take on auda-cious proportions. Because the city was so flat, its existing sewer lines simply filled with stagnant water during drier periods. When rain fell and the river rose, the sewers flowed backward into alleys, cesspools, and gutters throughout the city. Combining the sewer and sanitary lines into a water carriage system would only magnify the problem, causing sanitary sewage to back up into the city along with the stormwater.

Chesbrough's insight was to make Chicago less flat. He recog-nized that the water level of Lake Michigan was immovable, so that everything else would have to adjust to the lake. That is, rather than starting with the buildings and streets as fixed points, he solved the slope problem by working in the opposite direction. He started at the level of the lake and calculated the precise slope that would be neces-sary to keep the sewer lines flowing.

The new system involved using brick sewer mains from 3 to 6 feet in diameter, placed aboveground and running down the cen-ter of the street and into the collection point of the Chicago River. Because the city already existed at the older and flatter, inconvenient topography, entire swaths of the city had to be razed and then set at a higher elevation—literally, raised. The streets and the sewers of a city are its skeleton; the rest of the city infrastructure—buildings, hotels, and sidewalks—are the flesh and sinews that have to connect to this skeleton. As construction progressed away from the river, the sewer lines were raised to maintain the slope required to keep the water flowing in the pipes. Earth was then backfilled around the mains, elevating the street level along with the sewer mains. Close to the Chicago River, which drained into Lake Michigan, this plan required raising streets by just a foot or so; but farther away from the river, street levels had to be raised by as much as 10 feet to maintain sufficient slope.[12]

For a city whose skeleton was changed, adapting the remain-ing structures was a staggering achievement. Empty lots, created by tearing down buildings, were filled with enormous quantities

Laborers raising Briggs House, a Chicago hotel, in 1857.

of earth to raise them to the new street level. Many older buildings were left at the old level, creating a city that operated at two levels. For existing buildings whose owners wanted them at the new level—particularly brick buildings—raising them was a novel undertaking. In this era of entrepreneurial engineering, George Pullman—who would later become known for his palatial railcars—devised a method of placing the foundation of a building on a series of screws and then, by simultaneously turning all of the screws, lifting the entire building to the new elevation, where it was then supported by backfilling with additional earth. With hundreds of men turning the jacks simultaneously underneath each structure, building after building went through this topographic correction.

Chicago's topographic predicament is a lesson in the influence that sewage has over cities: buildings and streets can be corrected and adjusted; power lines or phone lines can be sagged and drooped and meandered about sprawling suburbs. But a pipe carrying sewage is an uncompromising necessity with particular topographic demands. By adjusting the topography, Chesbrough had managed to dry out the city. But in doing so, he exacerbated another problem. All of the sewage from the city and its factories was now carried that much more efficiently into the Chicago River and on to Lake Michigan, the city's source of drinking water.

Most early American cities sat on a major river or an estuary. Residents drew their drinking water from upstream and dumped wastewater downstream. A city on a lake presented a novel problem, for the source of water also had to be its toilet. This issue was something that Chesbrough, among many others, had long been wary of—and for good reason. In 1851 Chicago had lost 5 percent of its residents during a cholera outbreak, which was the impetus for hiring Chesbrough. But the contamination problem remained, even after the new sewer system was completed: in 1885, when over five inches of rain fell in 24 hours, rapidly flushing the city's waste into the lake, the resulting outbreak of waterborne diseases killed an estimated 12 percent of Chicago's population. Other cities had long since constrained waterborne disease; the highest typhoid rate between 1860 and 1900 was 4.7 per 10,000 inhabitants in a single year in New York, 6.4 in Philadelphia, and 8.6 in Boston. In Chicago, the rate exceeded 10 per 10,000 residents in eight different years over that period. Quite simply, Chicago's water and waste disposal systems made it a dangerous place to live. The solution that had once been financially unthinkable began to seem unavoidable.[13]

In the summer, rivers are slippery. Fly-fishermen often use felt-soled boots with their waders to help them stay upright and somewhat graceful as they scramble over riverbeds covered with "schmutz." The schmutz—a slimy, filmy substance that builds up on the rocks and gravel of streams and rivers—is critically important to the ecology of rivers. Most scientists don't call it schmutz, preferring instead the technical terms biofilm, periphyton, or even, drawing on the German roots of river science, aufwuchs ("overgrowth"). Regardless of what it's called, schmutz is a gumbo of bacteria, algae, and viruses as well as bits of dead plants. These organisms, the things they eat, and the things they excrete make up an ecosystem within an ecosystem, the study of which is now known as microbial ecology. This micro-ecosystem—or microbial community—also exists in the moving water of the river. A tiny fragment of a dead leaf, or even a particle of sediment, quickly acquires a community of bacteria determined to squeeze every last bit of food from its surface as it is whisked downstream. Most of the ecological work that takes place in fresh water is done by bacteria, and as long as there is oxygen in the water, bacteria will be there to consume organic pollution—including human waste.

The existence of this micro-ecosystem means that rivers, streams, and creeks are not sterile pipes that convey whatever is delivered to them to its destination in exactly the same form. Rather, anything put into a river is processed by the microbial ecosystem that lives in that river. Rivers are self-purifying. At least, that's what William Sedgwick claimed.

If Chesbrough, with his expertise in hydraulics and conveying water, represented the epitome of the mid-nineteenth-century water engineer, then Sedgwick was the consummate twentieth-century water engineer. Sedgwick spent the bulk of his career on the banks of the Merrimack River, which in the late nineteenth century was one of America's most developed—and consequently most polluted—rivers. It flowed through the mills and mill towns of the New England textile empire. All the waste and sewage generated by that textile empire was funneled into Lawrence, Massachusetts, whose geographic curse was being the most downstream town on the Merrimack.

At the turn of the twentieth century, a small research station was established in Lawrence right along the Merrimack River to study public health, especially the link between water and microbes. This water research station was run jointly by Harvard and MIT and was the precursor to what eventually became Harvard's School of Public Health. As its director, Sedgwick became the patriarch of modern water engineering.

Sedgwick initially pursued medicine at Yale; but after a few years, he shifted his focus toward the infant field of microbiology and finished his PhD, rather than an MD, at Johns Hopkins. Sedgwick was part of the bacteriological revolution of the late nineteenth century, an era that witnessed enormous breakthroughs as the links were finally made between contaminated water and disease. Before the work of these scientists, many diseases were attributed to miasma theory, the idea that putrefying wastes emitted pollutants into the air. Thus, foul smells were indicators of potential contamination. Disease outbreaks in cities were blamed on air rather than water, a partial reason for the focus on open boulevards and urban parks in early twentieth-century city planning. However, in the early 1890s a group of scientists and engineers led by Sedgwick scientifically confirmed the relationship between sewage-contaminated waters and typhoid fever, the great scourge of cities.

From his perch at the water experiment station in Massachusetts, Sedgwick became the sage of water, promoting its link to public health and the need for civil engineers to think as much about microbiology as about drainage. He also advocated for the treatment of water to avoid disease. Fortunately, in addition to identifying the impact of polluted rivers on public health, researchers of this time provided the means to clean river water: filtration and chlorination. With these methods, any city had invaluable tools to radically improve its public health by treating contaminated incoming water to remove the disease-causing microbes.

In addition to the direct treatment of water, Sedgwick also advanced the idea that running water was "self-purifying," or, as captured by a group of scientists: "Any stream will, if given time enough, that is to say, length enough, practically purify itself after receiving a given amount of sewage." Indeed, prominent texts of the time on water engineering, along with the leading research journals, began linking the natural processes of streams and rivers with water purification. Rivers were praised for their ability, with time and space, to take care of public health problems as well: "A few years ago it was stoutly denied that rivers had the power of purifying themselves. Then we knew practically nothing of nature's method. Now, that this has been so far revealed to us, it is declared with equal force that not only are effete matters rendered innocuous, but even disease-producing microbes are themselves voraciously devoured by others of like kind, and the formerly much-dreaded bacteria are—and properly so—considered amongst the best friends of man."[14]

Once Sedgwick and the growing water science and engineering community had shown that the combination of river self-purification and contemporary water filtration and chlorination was effective in removing disease-causing bacteria, they had established the justification to continue polluting rivers without restraint. As they argued, pollution of rivers would eventually be self-purified, and so society should harness the natural services provided by running water. Moreover, cities could always—and should always—treat the water they took in and distributed. As a result of these attitudes based on new scientific and technological discoveries, the most prominent water engineers of the era were justifying the practice of cities dumping their waste directly into streams and rivers untreated—as long as water was treated when used for supply, as quickly became the standard by necessity.

Sedgwick's philosophy of relying on river self-purification and water supply treatment was gospel to engineers but anathema to physicians. The use of self-purification to justify pollution marked a deep ideological divide between those like Sedgwick—who would become known as sanitarians or sanitary engineers and, eventually, environmental engineers—and those who had to deal with the consequences of sewage: the physicians and public health officials. The debate was not over what caused the diseases, but rather over how much financial cost cities should be expected to bear for water treatment.

Waste treatment before disposal would impose considerable costs on an upstream city to the benefit of a downstream city; and at the time few, if any, regulations required such treatment, let alone funding available to subsidize it. Moreover, because most cities had combined storm sewers with sanitary sewers in the model of Chesbrough's water carriage system, the volume of wastewater that would need to be treated was enormous—well beyond what could be handled by even the most aggressive of existing sewage treatment works. The presumption was that all cities would treat water at their point of use. If cities also treated their sanitary sewage before dumping it into rivers, then rivers and streams would be relatively unpolluted and cheaper to treat when used as source water. If instead cities were responsible for treating their water only at the point of use, then rivers would be much more widely and severely contaminated, and each water supplier would bear heavier costs to improve the quality sufficiently for safe use. One path led to some preservation of water quality in streams and rivers; the other led to vast pollution as every city primarily took care of itself.

Physicians have always been highly regarded in society, and municipal boards of public health had almost uniformly placed the task of water quality regulation under physicians. Yet, at the turn of the twentieth century, engineers were gaining a level of respect that challenged even that of physicians. In contrast to physicians, whose work in this era tended to be patient by patient, engineers in the early twentieth century were building the modern wonders of the world: monumental dams, sprawling road networks, complex bridges, and towering skyscrapers. Through their vast improvements to infrastructure systems, engineers were transforming society in grand leaps.

Sanitary engineers sought to do for sewage what their counterparts were doing with dams and bridges. These new engineers were biologists and hydrologists by training but planners and builders by practice. They were well versed not just in cholera and typhoid but also in landscapes, pipes, streets, and, crucially, budgets. As sewers and water supply systems became the skeleton and sinews of the modern urban landscape, engineers took on the expansive task of urban planning and, eventually, the role of city managers. They leveraged this role to establish their view of rivers as grand sewers of convenience for society. The argument proved compelling to other city officials, just as Chesbrough's appreciation for fiscal constraints had engendered enormous trust in his unusual plans to raze and then raise Chicago. As planners and city managers, engineers were involved in the fiscal realities and responsibilities of urban planning.[15]

Based on their role of solving both technical and fiscal problems for city governments, sanitary engineers argued that cities were responsible only for treating incoming water to provide clean water for their residents; they were not responsible for treating the sewage. Earlier engineering decisions set in motion by Chesbrough to combine storm and sanitary sewers now made sewage treatment all but impossible economically due to the sheer volume of water needing to be treated. The engineering community stubbornly took the line that it was not only preferable, but—as expressed in an editorial in the *Engineering Record* in 1903—"more equitable" for cities to be responsible only for water treatment at the intake and that they be allowed to discharge sewage into a stream untreated.

Their mixed mandate for public health *and* fiscal health shielded engineers from having to protect or be concerned with the need for sewage treatment before its disposal in streams and rivers. In a particularly rosy self-evaluation, a 1912 editorial in *Engineering News* boasted that the sanitary engineer was "a true and the greatest of conservationists, zealous to safeguard health and prolong life, but sparing no pain to see that each dollar is spent to the best advantage." Engineers used their awareness of fiscal constraints as a trump card to place their insight above that of idealistic physicians; they argued that it was engineers who had the unique and superior conception of the relative needs and values of municipalities, and they lumped together the entire medical community as "sentimentalists." In the end, the engineers won the battle, if not the war: in the first decade of the twentieth century, almost 90 percent of total collected wastewater ran into streams and lakes as raw, untreated sewage.[16]

All these aspects of that time—Chicago's flat topography, the science of self-cleansing rivers, and the acceptance of cities dumping pollution untreated into rivers—combined to bring about Chicago's second great topographic twist. While cities on a river could simply take water from upstream, treat it, and dump the waste downstream, Chicago was using Lake Michigan as both its faucet and its toilet. The only potential long-term solutions were to treat water as it became increasingly foul or continue to move Chicago's water intake farther out into the lake, beyond the reach of the polluting river. The city certainly attempted the latter plan: in addition to following Chesbrough's drainage and sewer plan, it also followed his instructions to build a new water intake two miles farther out into the lake, outside what was thought to be the reach of contamination. From this extensive distance, water was piped into the city through a series of over 240 miles of water pipes laid throughout the city. But even then, the problem remained that Chicago was producing enormous amounts of waste: by 1860 its population exceeded 100,000 people, and its stockyards were slaughtering almost two thousand head of cattle, sheep, and hogs daily. The wastes of this meat-packing empire flowed inevitably to the river, which became the city's festering open bowel. Or, as described in 1880, the South Branch of the Chicago River was "in an abominable condition of filth beyond the power of the pen to describe."[17]

Large rainstorms could flush out the river more fully into the lake, potentially reaching the water supply intakes. And so, the city decided to use its geography to separate the pollution from its water supply. The city would adopt what had seemed only three decades earlier to be an impossible plan; it would make the Chicago River flow backward.

Reversing the flow of the Chicago River was possible only because of the region's once problematic flatness. In fact, in its initial recommendation to investigate the river reversal, a commission noted that it was "practicable to restore the ancient outlet of the Great Lakes by opening a channel across the Chicago Divide, thereby creating a waterway to the Gulf of Mexico." The plan was simply to restore the ancient flow path. The eventual result was a 28-mile-long channel that served as both a sanitation drain for the city and an expanded navigation channel. The new channel, known as the Chicago Sanitary and Ship Canal, ran parallel to the older Illinois and Michigan steamboat canal. It was massive in comparison to the old one and substantially deeper, allowing passage of much larger boats than the previous canal could have accommodated. More to the point, the same qualities that enhanced navigation were also designed to pull water from Lake Michigan, so that Lake Michigan was no longer the

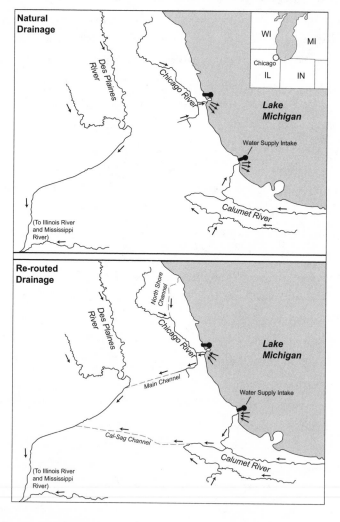

The Chicago River, before and after it was reversed to drain into the Mississippi River.

outlet for the Chicago River but rather the headwater. The Cal-Sag Channel to the south similarly flipped the flow of the Calumet River away from the lake and joined the main canal flowing away from Chicago.[18]

By pulling a continual flow of water in from the lake, the backward-flowing Chicago River allowed the fetid water to drain out and used the relatively clean water that remained to flush the pollution from tributaries, which it had now also diverted away from the lake. The engineers' critical concern was ensuring that the volume of water drawn from the lake would always be high enough to force the river backward—even during rain events. The initially designed and implemented flow that passed through the main channel was 10,000 cubic feet per second (cfs), just a bit less than the flow of the Colorado River as it passes through the Grand Canyon. The reversal worked exactly as planned for Chicago, but not for the downstream cities toward which its pollution was now careening: the switch drove the former Chicago River waters out of their normal watershed region into the Des Plaines River, which connected to the Illinois and Mississippi Rivers. As the historian Louis Cain describes the resulting revised geography, "Chicago was using the Great Lakes drainage area for its water supply, and the Mississippi River drainage area for its sewage disposal."[19]

This hydrologic coup d'état was politically possible because of the convoluted protocol for pollution treatment being championed by engineers. Sedgwick's vision of self-purification was crucial in determining how the newly downstream communities responded to the shift, now that all of Chicago's wastes were being swept through central Illinois and eventually on to St. Louis. Towns along the Illinois River, like Joliet and Peoria, were glad for the additional water to keep the river navigable during dry periods and for the subsequent increase in river traffic. But St. Louis, the other primary city of the Midwest, was less cheerful because the increased flow corresponded to what St. Louis thought was going to be dramatically increased pollution. Yet the river was in fact able to process the wastes along its course. Due to the length of the river flow and the constant oxygen-ation of the water, the microbial community could process the vast wastes in the river before it even reached Peoria, Illinois, much less St. Louis, Missouri.

In summarizing a vast array of studies on the subject, a series of academic scientists concluded: "It is evident that the Illinois River is capable of purifying itself to a very marked degree. The organic evidences of the Chicago sewage, as well as that introduced between Chicago and Peoria, have disappeared at Peoria, and we find the Illinois River at this point in as good a condition as the tributary streams."[20] Essentially, Chicago's adoption of Sedgwick's ideas about the self-purifying capabilities of rivers enabled the city to take the drastic step of reversing the river, and the same belief among the downstream communities made them willing, for a time, to accept it.

Though the ecological costs were thus easily dismissed, it was harder to stomach the colossal financial toll. Re-plumbing a city required digging up and rebuilding entire city streets—a huge burden when added to the growing list of responsibilities and costs borne by city governments. As more people moved into Chicago, as with any city, the infrastructure had to be scaled up and extended farther from the city center as overcrowded residents sprawled outward. And Chi-cago's government needed to fund those developments in addition to the costs associated with reversing the Chicago River.

The switch from the privy vault system to the water carriage sys-tem, along with the switch from water cart or private well to piped-in water, shifted urban water systems from a labor-intensive approach toward a capital-intensive approach. Privies had required unskilled laborers working constantly at what must have been a horrendously nauseating task. Though the technically complex water carriage sys-tem would be largely self-operating once complete, it required plan-ners, engineers, and enormous construction crews to design and build it. Water sources had to be identified, pumps built, and water mains constructed. Sewer systems had to be planned along topographic gradients that kept the water and waste moving continuously until it was eventually disposed of.

All of the houses and businesses had to be connected to the sewer lines. The route from the water source to the city, and from the city to the "toilet" of the waste stream, had to be surveyed by engineers, geologists, and planners. It had to be laid out in ways that balanced hydrologic and political realities of private property and myriad political subdivisions along the route of pipes and canals it had to cross. As such complications mounted, the costs of projects regularly exceeded municipal revenue many times over. These enormous costs came at the turn of the twentieth century, when the financial structure of the United States was undergoing its second great reordering.

From the founding of the United States through the Panic of 1837, state governments were the fiscal center of revenue and finance. After 1837 the financially crippled state governments had fiscally contracted, leaving their citizens' underserved needs to be filled by local governments. Consequently, the budgets of local governments grew dramatically in terms of the debt they took on to finance new projects as well as the revenue they took in to service this mounting debt. The growing debt took the form of the increasingly ubiquitous municipal bond.

As cities grew, and their revenue grew, "muni bonds" were seen as stable investments. When representatives of Chicago went to New York City to sell $400,000 worth of municipal bonds to finance Chesbrough's proposed sewer system along with other necessary water supply infrastructure, the bonds sold so quickly and were in such demand that representatives were able to sell a second round of bonds at far more favorable rates for the city. Throughout the late nineteenth and early twentieth centuries, Chicago's water and sewer bonds sold at interest rates comparable to those for American railroad bonds—one of the most stable private investments available. The City of Chicago and these new municipal water bonds in general were being seen as reliable, long-term investments within the broader investment markets.[21]

This fiscal soundness was not limited to Chicago: in the Panic of 1837, many state governments were unable to cover their debt; but during the Panic of 1893, which struck at the zenith of municipal fiscal growth, municipalities came out showing robust financial stability and certainly fared much better than the vast majority of private businesses. From the Panic of 1893 on, municipal bonds were considered extremely stable investments; boring, but steady. This growth was beyond what was occurring in the state and federal governments, which had contracted by a comparable amount. By 1902 local government tax revenues exceeded state revenues by 260 percent and national government revenues by almost 40 percent. Cities were now the fiscal heavyweights of governments.[22]

In 1905, waterworks—water supply and wastewater systems—had become the largest line item in the debts of municipal governments. In fact, waterworks costs often exceeded total revenue of even the largest cities: Chicago's waterworks in 1915 was valued at over $350 million, and yet the city's total municipal revenue was only $206 million: a 1.7 ratio of water cost to revenue. The ratio for Los Angeles was 1.48; for Harrisburg, Pennsylvania, 1.68; and for Ft. Worth, Texas, 2.32. Normally, such rapid growth in demand for debt would cause a corresponding increase in interest rates. Yet the growth in demand for municipal debt was not matched by high or rising bond yields. Bizarrely, municipal bonds continued to be characterized by relatively low interest rates amid ever-growing municipal indebtedness. Economists David Cutler and Grant Miller have suggested that this rise in municipal debt and lack of a corresponding rise in interest rates at the turn of the twentieth century was made possible by the development of increasingly sophisticated financial instruments. That is, the growth of the financial sector—of Wall Street itself—was part and parcel of facilitating the nation's municipal water infrastructure.[23]

The final monetary revolution devised by Chicago's water and sewer engineers, and eventually adopted by other cities, was an obscure but critical administrative and fiscal development: the special district. In Chicago's case, the special district, originally named the Sanitary District of Chicago, has been known since 1955 as the Metropolitan Sanitary District of Greater Chicago. Special districts were essentially an intermediate level of government in terms of size: they were under states but often spanned several municipalities. What made special districts useful was their fiscal and administrative independence from government entities. While municipalities were responsible for a range of functions within political boundaries from garbage removal to police forces, special districts could focus on specific services and had the flexibility of covering more regions on the basis of what made the most sense for each service.

In Chicago, city officials soon realized that the hydrologic landscape did not conform to the political boundaries of the city; the suburbs and newly connected upstream regions, which were part of the sewer system, extended well beyond Chicago's municipal limits. For the city, these suburbs were part of the problem because their existing drainage and sewage systems continued to route their waste directly into Lake Michigan. Thus, if the sewer system were constrained to the city of Chicago itself, the lake would still be polluted as long as the suburbs continued on their own independent paths. Residents of these outlying communities did not want to become part of the City of Chicago proper, but they did want to be part of the new drainage district. The use of a malleable special district allowed a broader geography of water and sewage service for suburbs of Chicago.[24]

Special districts also led to the implementation of new methods of financing government services, marking a continued fiscal evolution in the United States. In the early nineteenth century, state governments had relied on asset financing for revenues. That is, state governments financed infrastructure and then used the tolls paid by users of that infrastructure to service the debt incurred in building the infrastructure as well as cover the costs of daily operations. Meanwhile, municipal governments relied on property taxes, and the national government continued to rely on tariffs. Because local city governments bore the burden of providing most services in the early twentieth century, by 1902, property taxes accounted for 42 percent of all government revenues (national, state, and local levels combined).[25]

Following the innovation of the sanitary district in Chicago, the financial and administrative structure of the water and sewer district became a standard form of governance in the twentieth century. Other special districts formed, first in Illinois and then throughout the United States. The political and fiscal model of special districts, first established on the banks of a backward-flowing river, is the reason most city dwellers today pay their city a monthly water and sewer bill, based on the amount of water used, separate from taxes paid directly to the city.

CHAPTER 8

Burning Rivers

When a toilet is flushed in Durham, North Carolina, John Dodson has to deal with it. Dodson is a gentle giant of a man; tall and broad, but unassuming and patient. Along with wastewater treatment plant operators across the United States, Dodson is always on call, just like a physician with a pager. But when John gets a call from work, the issue involves tens of thousands of people and their unspoken millions of gallons of sewage.

On an average day Dodson supervises the treatment and processing of about half of the wastewater produced in Durham, a city of about a quarter million people. Each day over 8 million gallons of sewage come to his plant—about twelve Olympic-sized swimming pools per day of human waste. Across America others like John Dodson are managing the unseen treatment of human waste: almost 15,000 wastewater treatment plants pepper the country, along with over 700,000 miles of public sewer mains. All of this infrastructure is effectively invisible to the public, assumed to be working sufficiently well to justify being blithely ignored.

The waste from any toilet in Durham is transported through the pipes from the building where it's housed into larger collection pipes that combine with the many storm sewers of the city. Then it moves on to the sewage main—the large intestine of the city—where, along with the waste in thousands of miles of other pipes connected to countless other homes, buildings, and businesses, the collective waste is routed through ever more miles of pipe and pushed through the pipe system using sixty-six pumping stations spread through the city. Half of this waste eventually reaches Dodson's treatment plant, which sits innocuously at the edge of town alongside Ellerbe Creek.

The modern wastewater treatment plant is an extension of the work of early twentieth-century sanitary engineers: waste can be naturally processed, given enough time and the right conditions. The right conditions inevitably mean the right microbes, and Dodson is fixated on "the organisms," as he calls them, for his job is to ensure that the organisms have enough time and the right environment to do their work. When Sedgwick described streams and rivers as "self-purifying," he was really saying that they naturally contain the necessary microbes to process the waste.

In the early twentieth century, when streams were relied on to purify sewage, it was the distance between cities that made this approach effective. In the time it took for sewage to move from one city to the next along the streams and rivers, the microbes could process the waste. However, as America's population grew and cities expanded, the distances between communities decreased while the amount of sewage being loaded into the streams increased.

Modern treatment plants use the same fundamental processes to do the work of self-purifying rivers, but they have to do it in much less space. To accomplish this, modern sewage treatment plants use immense tanks that are constantly filling with wastewater. Some treatment plants serving small communities have tanks the size of home swimming pools. Others, like Dodson's, serve tens of thousands of people and have tanks that look more like a series of Olympic-sized pools: accompanied by the noise of pumps and blowers, wastewater constantly moves from one pool to the other, with pipes here and there moving wastewater from one tank to another. As noted earlier, the key to sewage treatment is time. Self-cleaning streams gained time thanks to moving sewage over long distances; treatment plants create time by slowing down the wastewater almost to a standstill. Dodson's tanks can slow down the millions of gallons a day and pass it through the plant in just under 24 hours, just enough time for the microbes to do their consumptive work. But the flow can only get so slow, because the tanks run close to capacity to save space—and all the while, more sewage is coming from the city as people keep washing dishes and flushing toilets.

Understanding what actually happens in a treatment plant takes a little microbiology. The main substance to be handled in sewage is the poop—alternatively, and perhaps preferably, called sludge. Besides sludge there are dissolved pollutants, primarily nitrogen and phosphorus, that environmental regulations require treatment plants to remove. Any substance or chemical is pollution only when it is present in excess; and though the large quantities of sludge and nutrients are pollution in streams, they are a feast for these "purifying" microbes.

Water treatment engineers have learned over decades that there are infinite varieties of microbes, each with its own peculiarity. Some microbes are like humans in that they consume oxygen and give off—respire—carbon dioxide. Other microbes are anaerobic; they live by combining sludge with carbon dioxide and respire methane. Still other microbes can use nitrate—a form of nitrogen—instead of oxygen or carbon dioxide.

These differences in microbes are what John Dodson must take into account. When he changes the flow or oxygen levels in the tanks, he will inevitably change the types of microbes that are doing the sludge eating. The design and engineering of sewage treatment plants is complex because it must provide a variety of ecosystems optimized for different microbes to eat the sewage. In natural streams, these various mini-ecosystem conditions exist in different places along the path downstream: oxygen-rich riffles and rapids provide habitat for some microbes while oxygen-depleted sandbars or stagnant pools are good for others. In wastewater treatment plants, the different habitats must be created at very large scales. When the different ecosystems are sequenced in the proper order, all of Durham's sewage can be converted into specific by-products—whether carbon dioxide or methane—between the time the water enters the plant and the time it exits and runs into Ellerbe Creek.

When wastewater comes into Dodson's plant from the sewer main, it is a grimy and smelly rivulet. As it enters the plant, the water passes through a screen to catch the large fragments of things in the water that microbes can't digest. This first step is simply to "get the crud out." After passing through the screen, the water—still viscous and rank—goes on to primary treatment: an enormous circular tank that looks and smells exactly like you might expect a sewage treatment plant to look and smell. This is where the initially foul work of treatment gets done. Each primary treatment tank is the size of a very large swimming pool and has a slightly cone-shaped bottom, where the sludge settles while the grease floats to the top. Scrapers and skimmers—long mechanical arms—do the work here: slow-moving scrapers creep along the bottom of the tank to force the settling sludge toward a collection hole at the bottom while skimmers at the top push the floating grease into collection troughs, from which it is first routed to a dumpster and then hauled off to a landfill.

The sludge has a more interesting fate. For most of the sludge, its next stop is a smaller circular tank with an enormous lid on top—an anaerobic digester. This little ecosystem has no atmospheric oxygen, and so the only sludge-eating microbes that can grow are the ones that convert the sludge into very small amounts of solids and very large amounts of methane, or natural gas.[1] The methane is stored in an enormous gas tank and will be used for heating buildings. From the top of the three-story-tall gas tank, Dodson points to an enormous covered area a few hundred yards away, where bulldozers are busily pushing stuff around. That is where they dry the remaining biosolids into what is effectively concentrated fertilizer. Every day, dump trucks haul away seven or eight loads of concentrated biosolids to be delivered for use on over three thousand acres of farmland.

Pointing out over the acres and acres of tanks, pools, pipes, and trucks, Dodson says that settling the solids is the easy part; the hard part is treating the dissolved nutrients. Anything dissolved in the water moves right along past the settling tank, like sugar dissolved in water. The two main soluble headaches in Dodson's life are nitrogen and phosphorus. Despite the challenge of processing these two elements, environmental regulations require them to be reduced as part of wastewater treatment because of their effect on downstream ecosystems. This processing is done in the secondary treatment tanks.

Water coming to secondary treatment tanks is still gray-brown and foamy, laden with nitrogen and phosphorus. Whereas primary treatment was one big circular tank, secondary treatment takes the form of a very long rectangular tank broken up into a series of smaller, but still massive, sections. Each of the smaller sections is the size of a two-story house, and each one is sized specifically to ensure that the water spends the necessary amount of time in that section of the tank. Imagine a series of two-story houses with no roofs, lined up next to each other, each with a single open window in the second story. Water fills the house and pours from one house to the next through that single open window. To the naked eye, each sectioned-off tank looks the same as all the others, but every tank contains its own particular ecosystem with its own particular processes: one causes phosphorus to be released from the suspended sediment and become soluble in the water; another contains microbes that convert ammonia to other forms of nitrogen; another contains microbes that drive nitrogen out of the water and move it inertly into the atmosphere, and so on. Careful sequencing of conditions for microbes is the heart of wastewater treatment.

All of these microbes in secondary treatment must eat, and this need highlights one of the more interesting tricks of the modern sewage treatment plant. Some of the sludge from the primary treatment is mixed with incoming sewage and sent to this series of secondary tanks to feed the microbes. By the end of these secondary aeration tanks, the microbes are essentially out of food; they've eaten everything available to them, and the amounts of phosphorus and nitrogen dissolved in the water have dropped dramatically.

Even at this point, the water looks muddy and turbid; not as viscous, soupy, and frightening as when we started, but not something you would want to swim in, either. The next stop in the plant is where things visibly change. The wastewater flows into a second set of circular tanks, where the microbes come together and settle, leaving the "clean" water on the top. The water is then trickled through another filter: a bed of gravel, sand, and coal. When it comes out, it looks downright drinkable—but it isn't, quite. From the filter, the water is funneled through a narrow channel that passes under bright ultraviolet lights to neutralize any remaining pathogens. Finally, it flows down a ditch that exits the treatment plant and leads into Ellerbe Creek. The contrast between the water coming into the plant and the water coming out is often stunning. In fact the water coming out of the plant is noticeably cleaner than the water in Ellerbe Creek, where Dodson points just a bit downstream to indicate some bass swimming in the output of his treatment plant.

Ellerbe Creek feeds into Falls Lake, which, while not as polluted as Lake Michigan at the turn of the twentieth century, is nonetheless the recipient of the wastewater of several towns and cities along with hundreds of farms. In ongoing efforts to clean up the lake, the Environmental Protection Agency (EPA) has made life more and more difficult for the operators at Durham's wastewater treatment plant. Not all of the nitrogen and phosphorus can be removed from the water as it passes through treatment plants, and Dodson has to monitor how much nitrogen gets into Ellerbe Creek as part of his day-to-day operations. In 2014 the total nitrogen was about 103,000 pounds. In 2011, the EPA and the State of North Carolina set a sequenced series of limits to how much nitrogen could be released into Ellerbe Creek; Phase 1 limits were set at 97,000 pounds per year. "You know, we can hit that," Dodson said. "I'll have to push the plant a bit more, but I can squeeze things down to 97."

Phase 1, however, implies a Phase 2, and that is where Dodson foresees trouble. Phase 2 would limit releases of nitrogen to only 67,000 pounds per year—a 30 percent reduction that would be beyond the technology used to operate his plant. "That would be a transformation. We'd have to move to reverse osmosis. And if we went to that, Durham would have drinking water coming out of their wastewater plant." The problem to Dodson is not that the water coming out would be too clean. The problem is identical to the problem that Chesbrough and Sedgwick faced over a century earlier: the costs would be enormous, but the residents wouldn't benefit. When Dodson reflects on the future, he just shakes his head: "If they go to Phase 2, I bet they bankrupt the city."

How did the federal government come to require cities to spend enormous amounts of money on sewage treatment? Like many others in American history, that story stretches back to the Great Depression and the World Wars.

The first iteration of John Dodson's North Durham Treatment Plant, like many treatment plants in the United States, was built during the Great Depression, the third great fiscal pivot of U.S. governance. If, from the founding of the United States until 1837, states had been the financial center of government and the Panic of 1837 pivoted the U.S. financial system to becoming city centered, then the Great Depression and World War II handed the financial reins to the federal government.

The seeds of this restructuring were sown in 1913 when Congress passed the Sixteenth Amendment, which allowed the federal government to begin collecting income taxes in 1914 to fund the costs of World War I. Before 1914, the federal government's share of taxation tended to rise during wars, primarily through tariffs until the start of World War I, but then decrease as war-related debts were repaid. But the close of World War I was different from all previous wars. Even after the war ended in 1918, income taxes remained in place for the federal government to continue generating revenue.[2]

A dramatic change in the location of government debt also took place during this era. The federal government was largely debt free in the early twentieth century whereas local governments had taken on substantial debt leading up to the eve of U.S. involvement in World War I, when municipal debt made up over 70 percent of total government debt. This situation reflected the central role that cities played in society by providing most of the services associated with government—water supply, firefighting, police, roads—while the federal government remained relatively inert. Local governments generated revenue primarily from property taxes. If you were alive in the early twentieth century, you sent most of your taxes to your local municipal government—and the size of your tax bill was determined by where you lived rather than how much money you made.

When the Great Depression hit, most people could no longer afford to pay their property taxes. The cities, unable to collect, were likewise unable to pay their existing debt. As the Depression deepened, cities could no longer justify taking on additional debt; so any infrastructure projects, even those already begun, ground to a halt for lack of funds. Cities struggled to sustain basic functions like waterworks, railroad projects, and police services. For cities that could not even pay for drinking-water infrastructure and services, sewage treatment works were far down the list of potential projects.

In response, the federal government filled the vacuum left by city governments with Franklin Delano Roosevelt's New Deal. Between the federal programs and projects of the New Deal and the costs of World War II, the federal government's portion of total government debt soared to over 50 percent by 1942, where it has remained ever since. The shift in spending and debt was accompanied by an equally dramatic shift in revenue: income taxes rose to offset the increased federal debt, increasing from around 20 percent of total government revenue in 1920 to over 50 percent in 1942. Property taxes, in contrast, plummeted to below 10 percent of total government revenue after the war. As the economic historian John Wallis would note in his analysis of fiscal policy, the United States emerged from World War II with a fundamentally different financial structure than the one it had entered the war with—and, in fact, a different structure from what the country had used since its founding.[3]

With this dramatic change in taxation came equally dramatic changes in how different levels of government got their money, and an increasingly convoluted revenue system. Before the Depression, there was general consensus that federal funds should be spent only on projects of national need, and local funds would be spent on projects of local interest. Property taxes would be linked to municipal projects, tariff taxes to national projects. But the New Deal created a fiscal system through which the federal government collected income taxes nationally and then spent that revenue via grants to state and local governments—an enormous redistribution of funds and government power across the country.

This is exactly how funds were obtained for Durham's first sewage treatment plants and Chicago's wastewater treatment plants of the early twentieth century. In Chicago's case, using the backward-flowing Chicago River proved to be only another short-term solution; the city ultimately had to construct a wastewater treatment plant capable of handling most of the city's wastewater. What's more, Chicago was hit hard by the Depression; by January 1, 1932, the sanitary district was unable to sell bonds—no one would loan the city money—to finance ongoing and proposed work, thus bringing its planned treatment plant to a standstill. To fill the financing gap, in 1933 some of the first grants given out by the Public Works Administration (PWA) were routed to Chicago. Rather than doing the work itself, the PWA loaned the city money by purchasing municipal bonds from Chicago to finance the wastewater treatment plant. The sewage treatment plants that resulted would eventually be the largest in the world.

The best-known public works construction feats of the New Deal are the large federal dams, Civilian Conservation Corps (CCC) stone bridges, national parks, and expanded highway system. Yet the program of federal government grants to local governments run by the PWA was no less influential. It was instrumental in the rapid expansion of sewer and wastewater treatment, supporting over 1,800 sewer projects across the nation. Between 1933 and 1939, the PWA financed almost 65 percent of the nation's sewage plants; in the late 1930s, it supported around 80 percent of the total new sewerage system construction nationwide, including that infant North Durham Plant in North Carolina that John Dodson would operate eight decades later.[4]

Just as the New Deal brought electricity to the rural masses, the change in fiscal order also brought sewerage to the urban masses: the urban population serviced by sewers rose to almost 75 percent by the close of World War II while urban population rose by almost 50 percent over the same time period. By the 1960s, over three-fourths of the nation's sewage was treated in some way before being routed into streams and rivers.[5]

Recovery of the national economy after World War II, however, did not lead to another shift in the fiscal roles of the various levels of government: federal aid to cities had replaced property taxes as the primary source of revenue for city services, and it would remain a primary source of revenue for several decades despite the financial recovery of cities across the United States. In 1956 the federal government provided 30 percent of costs for building sewage treatment works, and that number grew to 50 percent in 1966.[6]

Cleaning up the pollution from America's rivers was done in the shadow of FDR's Public Works Administration, accomplished as much through a philosophical change in government revenue and financing as it was through chemistry and engineering. And as the federal government took hold of the purse strings for these projects, it simultaneously introduced federal regulations for water quality. The capstone of this trajectory was the Clean Water Act of 1972, a federal legislative action inseparably associated with a burning river and the tireless work of an Ohio housewife.[7]

———

The middle of the twentieth century marked the point when America's rivers became not just polluted, but industrialized. The growth of oil refineries, mining, and particularly the organic chemical industry introduced to America's rivers an entirely new array of wastes, with largely unknown characteristics, leading to unforeseen consequences. Initially, industrial wastes were seen as innocuous or even beneficial; they were thought to have a bacteriological effect with the potential to kill the pathogenic cholera- and typhoid-causing microbes. But it gradually became clear that these new compounds—benzene, carbon tetrachloride, tetrachloroethane, and trichloroethylene, among a tongue-twisting list of many others—were more toxic and longer lived than pollutants of previous eras.

While the Monsanto Chemical Company during the 1940s separated its toxic waste to bury rather than dump into the Mississippi River at East St. Louis, most chemical manufacturers were not so progressive. Many industrial manufacturers began setting limits for how much time their own employees could be exposed to these chemicals, yet they continued to dispose of the compounds and wastes by spilling them untreated into streams and rivers. During the late 1930s, New Jersey investigators estimated that almost 7 million gallons of untreated industrial chemical waste were dumped each day into New Jersey's streams. This practice continued into the late 1960s, when only 15 percent of the chemical industry's waste received any treatment before being dumped into streams and rivers.[8]

The geographical clustering of polluting industries soon left its foul traces on the American landscape: chemical manufacturers in the Mid-Atlantic region left vast amounts of pollution in the river networks of New Jersey and eastern Pennsylvania; eastern Louisiana and western Mississippi became egregiously polluted from the petrochemicals dumped by the dozens of factories along the banks of the Lower Mississippi River; streams and rivers draining the copper-laden mountains of Montana became lunar landscapes of mine tailings. The Cuyahoga River was particularly extreme; it touched not only the coal and steel empires of the Ohio River valley but also the hub of John D. Rockefeller's Standard Oil empire of the Great Lakes. The banks of the Cuyahoga were already home to twenty refineries during the Civil War, and through the turn of the twentieth century, steel mills, chemical plants, paper mills, and all other manner of industrialization sprang up along the river.

On through the mid-twentieth century, the auto industry became integral to the Upper Midwest economy and proved particularly damaging to the health of the Cuyahoga. Although the cars were primarily assembled in Detroit, Michigan, many of the individual parts were manufactured along the Cuyahoga River in Cleveland and Kent, Ohio. Farther upstream on the river sat Akron, Ohio, the largest rubber-manufacturing city in the world. Waste flowing out of the wastewater treatment plant in Akron accounted for over two-thirds of the total river flow, making the Cuyahoga less a channel of water than a roaring stream of industrial waste. In 1968 the Cuyahoga was assessed by a group of scientists, who found that in Akron the river was a "grossly polluted zone," while at the most downstream end, just upstream of where it enters Lake Erie, the river was "grey, septic in the pools, and odorous," and noted for not meeting any "water quality criteria for any use."[9] The Cuyahoga had become an open sore on the landscape.

Edith Chase—the conservation pioneer who cleaned up the Cuyahoga decades ago—is now 89 years old. She is just a shade over five feet tall and has a shock of dark, undisciplined hair. She is mild mannered, but the Ohio Environmental Council calls her the "Grand Dame of Lake Erie." She is an unabashedly proud housewife who has accomplished more in that informal role than most achieve in a formal career at a regulatory agency: despite never having a formal policy career, Chase received a lifetime achievement award from the National Oceanic and Atmospheric Administration (NOAA). She has a simple philosophy for fixing the environment and changing the world: "You have to get mad, and you have to get lucky."[10]

In the 1960s, Edith Chase was a recently transplanted housewife in Kent, Ohio, which was then a bustling college and manufacturing town. But the town was also beginning to see the initial hints of decline. Factories were being downsized, and a few were shuttered. A smattering of houses had fallen into disrepair, and the river was in atrocious condition. With an undergraduate in chemistry from Antioch College and a masters in organic chemistry from the University of Minnesota, Chase was well aware of what was causing the local river to decline. She began to look for ways to get involved.

What she found was the League of Women Voters, a group of well-educated, well-networked, and well-organized women throughout the United States. Formed in 1920 in the aftermath of the women's suffrage movement, the league already had a long list of achievements, including being instrumental in passing the Tennessee Valley Authority Act (TVA), which they supported because it brought electricity to households—and women—throughout the Deep South. The league followed a long tradition of women playing a leading, yet often behind-the-scenes role in environmental conservation, ranging from the formation of the Audubon Society to the initiation of the first systematic street-cleaning programs. Women, and specifically housewives, could and often did base their conservation claims on the socially laudable grounds that their interests were in protecting their home and children. Water quality in particular became a central focus in the 1950s and 1960s: no issue was as critical to housewives as ensuring that their children could safely drink water. Water pollution quickly became a topic discussed by housewives and those catering to housewives—before long, Good Housekeeping magazine began carrying in-depth articles on water pollution in America's cities and suburbs.[11]

In the industrialized Upper Midwest, the League of Women Voters set its sights on Lake Erie, which was typically described as either dying or already dead. Lake Erie received over 750 million gallons a day of untreated—or barely treated—municipal waste and 2 billion gallons a day of industrial waste. In the 1950s the Niagara River, the outlet of Lake Erie, was proposed by state regulators to be reclassified from a Class A river to a Class C river—an official indication that the river was not fit for human consumption, bathing, or swimming. Tourists at Niagara Falls were basking in and spellbound by the mists of what was effectively a cascading open sewer. Tens of thousands of dead fish washed up on the banks of Lake Erie tributaries, and in 1965 a "dead zone," where the water quality and oxygen levels were so low that few organisms could inhabit it, was discovered in the middle of the lake. The zone spanned over 2,600 square miles—an area larger than the state of Delaware.

The League of Women Voters launched the political battle to clean up Lake Erie in 1963 with the formation of the Lake Erie Basin Committee. Edith Chase joined the league a year later and eventually became chair. Almost fifty years after working on the lake and the river cleanup, she can still rattle off chemical compounds and their decay rates as easily as she recites her grandchildren's birthdates. She cites the names of scientists as well as the dates of their relevant studies and publications. And she recalls the first and last names of the U.S. newspaper reporters she worked with to publicize the league's studies and findings. Just as impressive is her clear command of the policies and regulations; she knows those of the past and the present.

In her position as chair, Edith Chase played a central role in developing and implementing the league's strategy for cleaning up Lake Erie—a strategy that would serve as a model for any twenty-first-century environmental nongovernmental office (NGO). As she herself argued convincingly, "The League was way ahead of the game on water pollution." Armed with housewives like Edith Chase, many of whom lived around the lake and along its vile tributaries, the league dug into the topic, ultimately writing and publishing one of the most comprehensive and coherent reports on the Erie yet available: *Lake Erie, Requiem or Reprieve?* There was no shortage of scientific journal articles, technical reports, and op-eds on Lake Erie at the time, and the league had to compete with them to draw attention to potential solutions. Yet while most reports on water pollution were dense, towering, and focused on chemistry or pictures of dying fish, the league's report managed to be both sweeping in its scope and, at a slender fifty pages in large font, concise and to the point. It reviewed the geologic history of the basin, synthesized chemistry data, analyzed scientific studies, and then did what few others had done—it identified the disparate policies and regulations that might improve water quality and identified where the bottlenecks were for making existing policies work.[12]

One of the most insightful elements of the league's report was its recognition that political decisions would always come down to finance. In the (ironic) gendered convention of the time, the league wrote, "The citizen must reconcile himself to the inescapable cost of pollution abatement and realize that he will pay in all his roles." Cleaning up the environment needed to happen, and it wasn't going to be free. In addressing this line of thinking, the league raised the biggest issue standing between the activists and a solution: Who was going to pay? Specifically, they focused on what should be the division of costs among federal, state, and local budgets.[13]

Armed with the report, which drew considerable attention from Upper Midwest media and politicians, the League of Women Voters organized a series of clean water conferences. In turn, league members were invited to major national conferences. They incessantly pressured local and state politicians, which led the Cleveland legislature to pass a $100 million bond issue to finance river cleanup, sewer system improvements, stormwater overflow controls, harbor improvements, and debris removal. In short, the league had successfully convinced the general public in Cleveland and the surrounding communities to take on significant debt for cleaning up the river.

Despite this victory, there remained major barriers both political and financial. While the voters had approved going into debt to clean up the river, the city government was wary about actually selling the bonds. Additionally, money from the city's bonds would address only the public side of pollution—sewage—and cleanup efforts would be effective only if the industries treated their own wastes as well, at their own cost. State and local politicians were not willing or able to take on the polluting industries, because any threat of local regulation would simply drive away the companies that had converted Ohio from frontier backwater to industrial empire—and those firms employed many of the politicians' constituents. In fact, the state government added insult to injury by reclassifying the Cuyahoga as an industrial river, meaning that polluting facilities along the river were immune to public nuisance lawsuits and thus protected from local efforts to combat pollution.

The League of Women Voters had reached an impasse. For most of the 1960s, the league and Edith Chase had done an enormous amount of work in building coalitions and arguing persuasively that changes were necessary. But they hadn't caught a break.

Then things started to go the league's way in the summer of 1969—ironically, when the Cuyahoga caught fire, the *Apollo 11* astronauts returned successfully to Earth, and young Ted Kennedy drove his car off a bridge.

The Cuyahoga had caught fire at least a dozen times since the Civil War and probably many more than that. By the 1930s, the local press was more critical of the lack of firefighting services to put out the river fires than they were of the river burning in the first place. Throughout that period, other industrial rivers also routinely caught fire, including the Buffalo River in New York, the Schuylkill in Pennsylvania, and the Rouge River in Michigan. Compared to some of these earlier fires, the 1969 blaze on the Cuyahoga was small, easily contained, brief, and seemingly insignificant. It was so short lived, in fact, that local photographers were not able to get there in time for a photo. In the Cleveland papers, the fire was barely mentioned; it merited a single paragraph in one paper and back-page coverage in another.[14]

Local coverage never would have helped the league's push for calling attention to water quality. But only a few weeks after the Cuyahoga fire, the *Apollo 11* astronauts returned from the moon. Surprisingly, even this giant leap for mankind got secondary cov-

The infamous photograph of the burning Cuyahoga River in 1952, printed in the August 1, 1969, issue of *Time Magazine*.

erage in the news cycle thanks to the scandal at Chappaquiddick Island, where Senator Ted Kennedy had driven off a bridge. He escaped; but his passenger, Mary Jo Kopechne, drowned. In a single week, men returned from the moon and a famous senator was caught in a scandal. It was a magazine publisher's dream. Both were featured in the August 1, 1969, issue of *Time Magazine*, which splashed a photo of Ted Kennedy in a neck brace across the cover. And there, tucked away in that issue's science section just behind the series of articles about lunar exploration, sat an article about water pollution in cities. It ran with the now infamous image of the Cuyahoga burning: the photo was taken in 1952, yet it was attributed to the event in 1969.

In recounting that summer forty-five years later, Edith Chase said with a sly grin, "Poor Ted, but lucky us."

The aftermath was the stuff of legend. The burning Cuyahoga, regardless of the date, gave the league and the environmental movement its catalyzing image. People everywhere bought the magazine to see Ted Kennedy and the astronauts but then browsed their way to read about the burning river in Cleveland. Faced with the embarrassment of the river burning, the federal government and the State of Ohio sprang into action. Within weeks of the blaze, the federal government threatened six industrial firms—including two steel mills along the Cuyahoga—with prosecution if they did not reduce pollution. Over the next 6 months, another sixty-six federal prosecutions were commenced against industrial water polluters around the United States. And from there, in the frenzy of federal environmental activism, came the Clean Water Act, the Clean Air Act, and the National Environmental Policy Act—the environmentalist's holy trinity of regulations.

The Clean Water Act implemented standards for pollution control and reduction. It gave the newly created federal Environmental Protection Agency (EPA) teeth—where state agencies, like those in Ohio, had none—by giving it the authority to permit or deny water pollution. That is, if manufacturing plants or public wastewater plants wanted to route their wastewater into a stream or river, they had to ask the EPA for permission. And the EPA made treating that waste a condition of granting permission.

If the EPA's permitting program was a "stick" to force pollution reduction, the 1972 Clean Water Act also had an extremely tasty carrot: a generous federal spending program that funded initiatives to clean up the nation's rivers. Whereas previously cities like Cleveland had footed the bill for cleanup, now the federal government would provide up to 75 percent of the costs for planning and building wastewater treatment plants. Even FDR's New Deal, with its expansive building programs, had been minor compared to the spending spurred by the Clean Water Act. In the first twelve years after the act was passed, the federal government spent over $40 billion on wastewater treatment. By 1990 there were over 16,000 centralized public wastewater treatment facilities, servicing over 75 percent of the U.S. population, along with an additional 160,000 industrial treatment sites.[15]

The 1970s would prove to be a peak for both regulating water pollution and spending in a fiscal era. For just over six decades— from 1914 through the middle of the twentieth century—the federal government used income tax to increase its own revenue and then used that revenue to take on the central role in government spending. State and local governments, the predominant sources of revenue for infrastructure in the nineteenth and early twentieth centuries, gradually adapted their budgets to be dependent on funding from the federal government. In 1978 grants accounted for almost a third of state and local government revenues, and even more for some cities: that year, federal grants made up 55 percent of St. Louis's budget and 69 percent of Buffalo's. The federal government had become the primary taxing entity, redistributing as it saw fit to state and local governments, which gradually decreased their own taxes thanks to the availability of grants from the federal government. The cleanup of rivers polluted by industrial and municipal wastes was made possible as much by this mid-twentieth-century approach to fiscal governance as by the introduction of new regulations.[16]

In the 1980s, the United States would undergo its next dramatic financial and political reorganization with the Reagan Revolution. And once again, river pollution would be part and parcel of its effects.

For all the benefits of the Clean Water Act, it had an enormous blind spot: Farmers and suburbanites with Irish-green lawns were let off the hook. When a farmer fertilizes a field, or a homeowner uses a bag of fertilizer from the local garden store, they play a role in changing the chemistry of the planet.

The ability to grow plants is limited by the availability of a handful of elements, most notably nitrogen and phosphorus. The central goal of agriculture for centuries, and the real work of farming, has been in providing nitrogen and phosphorus. In search of nitrogen (though they didn't know it was nitrogen they were after), twelfth-century Japanese farmers brought "night soil"—human waste from outhouses—back from the cities they visited to sell crops; and nineteenth-century British farmers imported bones from other European countries to supplement their own depleted soils. Until the 1840s, the principal commercial nitrogen fertilizer was guano, the solidified bird excrement that accumulates on a few arid islands. Regular commercial shipments of guano to New York started in the 1840s and totaled about 760,000 tons in the 1850s. At that point, in 1856, Congress passed the U.S. Guano Act, declaring that any uncharted island containing guano could be claimed as the property of the first U.S. citizen to stumble upon it. For phosphorus, sources were available closer to home; commercial-scale deposits were found in the southern United States in the late 1800s and quickly grew to global importance during the early twentieth century.[17]

The game changer came in 1913, when two German scientists perfected the synthetic production of nitrogen fertilizer. This breakthrough occurred simultaneously with the development of commercial-scale phosphorus mining, such as those enterprises growing rapidly in Florida. Thanks to these new sources of nitrogen and phosphorus, for the first time in human history farmers had an effectively unlimited supply of the primary nutrients needed to fertilize plants. After realizing the increase in productivity made possible by using synthetic fertilizer, the farmers' conversion was rapid and thorough: by the 1960s, over 90 percent of all cornfields in the United States were fertilized with synthetic nitrogen. The amount used per acre increased by a factor of ten over the second half of the twentieth century, when farmers began applying more than was needed to ensure maximum crop yields. As farmers used more and more fertilizer, the plants were unable to absorb it all—and more of the nitrogen ran off the fields, into the irrigation ditches, and on to streams and rivers.[18]

By the 1980s, just under half of the nitrogen fertilizer applied in the Mississippi watershed region ended up in the Gulf of Mexico rather than in crops. Just as fertilizers instigate the growth of plants on land, they also instigate the growth of plants in water. When they wash off farmlands in the Midwest and eventually fertilize the tiny plants—algae—floating in the Gulf of Mexico, they create massive algal blooms. As the algae die, they sink to the bottom of the Gulf, where their tiny bodies feed booming numbers of bacteria that use up all the oxygen in the water. The resulting dead zone of low or even no oxygen in the Gulf now typically exceeds 7,700 square miles, making it just a bit smaller than New Jersey. Through this cascading ecological pathway, farming in Iowa is destroying shrimping in Louisiana.[19]

Farms were not alone, and the Gulf is not alone, because any activity using fertilizer leads to runoff to rivers, which delivers the excess fertilizer to whatever water body is downstream. Every suburbanite engaged in an endless quest for the greenest lawn in the neighborhood follows the same rationale as a farmer (over-fertilize), and the fertilizer from Home Depot has the same effect as the fertilizer used in commercial agriculture. Just as streams in the early twentieth century were a dumping ground for sewage, streams in the late twentieth century have become a commons for fertilizer. By the turn of the twenty-first century, the EPA had classified just under half of the 3 million river and stream miles in the United States as either threatened or impaired, and the cause was fertilizer from agricultural and suburban runoff.[20]

In cities and suburbs, the degrading aquatic effect of fertilizer is combined with the success of decades of water infrastructure construction. In many suburban areas, sewer lines carry both human sewage and storm flows; this is the legacy of Chesbrough's water carriage system, introduced a century and a half ago. However, by being combined, both sewage and stormwater must all go through a treatment plant, there to be handled by the John Dodsons of the United States. On a normal dry day in Durham, John Dodson's treatment plant has to manage about 8 million gallons of primarily sanitary wastewater. On a rainy day, about 40 million gallons will come rushing through the pipes, and the extra 32 million gallons comes from runoff—storm sewer. All of this water has to be processed; it has to be slowed down long enough for the microorganisms to do their work. And with all the fertilizer on lawns, the organisms have to consume that much more nitrogen and phosphorus.

On the rainy days, Dodson wonders if his treatment plant can handle all the water pouring through its system; someday, will there be just too much water? These are the days when problems arise, when wastewater treatment plants have to bypass treatment altogether. Because all of the sewage is combined, very rainy days mean that in addition to suburban runoff, untreated sewage bypasses the treatment plant altogether and flows directly into streams and rivers. This is the worst-case hydrologic scenario—pollutants and pathogens passing on to downstream water supplies along with untreated fertilizer and thus exacerbating dead zones in reservoirs and estuaries. John Dodson knows these untreated overflows can incur the regulatory wrath of state and federal regulators.

In Alabama the worst-case hydrologic scenario occurred simultaneously with the worst-case financial scenario, all made possible by the most recent financial transformation in the United States: the Reagan Revolution. The perfect storm occurred at the Cahaba River in Alabama, which went from anonymity to infamy in the world of finance.

The Cahaba River is a perfectly nice river. Contained entirely in Alabama, the Cahaba starts its 190-mile route near Birmingham, semicircles through the state, and culminates by joining the Alabama River just outside Selma. Amid the rapid suburbanization of the U.S. South in the mid-1990s, sewage treatment plants were becoming overwhelmed with runoff and sewage, and those along the upper Cahaba River were no different. Moderate rainstorms—common in the humid Southeastern United States—caused repeated overflows during which the wastewater, including raw sewage, bypassed the treatment plant and flowed directly into the river.

In addition to implementing new regulations and providing funding for wastewater infrastructure, the Clean Water Act also enabled citizens to sue when those regulations were being violated. In the mid-1990s, the Cahaba River Society did just that; it sued Jefferson County for violating the Clean Water Act. After some legal back-and-forth, in 1996 Jefferson County agreed to take on the task—and cost—of eliminating sewage overflows. The county planned to take over management of twenty-one different sanitary sewer systems along the river and improve the existing treatment plants. It would also improve over 3,100 miles of sewer lines, some of which were found to be made of clay installed over a century earlier. Initial cost estimates came in at a shade over $250 million, but they quickly ballooned to $1.2 billion.

In the 1970s, Jefferson County could have expected that substantial federal grants would be available to finance the work. However, the era of money flowing freely from Washington, D.C., to cities and states came to a screeching halt as Ronald Reagan's administration took hold in 1981. As the fiscal conservative in chief, Reagan reined in the regulatory and environmental spending portions of the federal budget—in order to cut income taxes—by pushing the responsibility for financing projects back down to more local levels. It worked: by 1991 the federal share of spending on wastewater infrastructure had dropped to 5 percent. This was a jarring reality for local governments, which suddenly faced the reality of paying for the requirements of the Clean Water Act on their own dime.[21]

Besides the Reagan Revolution, another dramatic financial transition was occurring in the world of finance: this was the era that introduced the world to the vocabulary of derivatives, options, and swaps. The best way to understand these abstract concepts—what they are and why they are used—is to look at how sewers in Jefferson County were financed.

For Jefferson County in the late 1990s, the costs for the revamped wastewater infrastructure system would be large, but the finance mechanisms to pay them were, at least initially, tried and true: muni bonds—a long and steady, even boring, form of investment due to their extended duration and nearly guaranteed returns. Municipal bonds are simply loans given to a city to do work, whether for building a hospital, public housing, or a sewage system. When an investor buys a $10,000 municipal bond from a city, that investor is giving the city a loan for $10,000. In return for the loan, the city pays the investor some percentage of that loan each year for some number of years. At the end of the term, the city returns the original total, $10,000. If you buy a 20-year, $10,000 bond at 5 percent interest, you will get $500 each year for 20 years; at the end of the twentieth year, you'll get your $10,000 back—as long as the city has the money to pay back all those loans. The city must then set its water rates high enough to generate revenue, which the city uses over time to pay back the interest and principal for all those bondholders. Because infrastructure is expensive as well as long lived, municipal bonds tend to be very large and long term; it is not unusual for municipal bonds for a sewer project to provide tens of millions of dollars in funding and have loan periods of 10 to 30 years.

Loaning a city money for three years is relatively low risk, and so the interest rate for a three-year municipal bond would typically be very low, perhaps 1 percent. Loaning a city money for thirty years is riskier; lots of things can change in a city over three decades. Therefore, the interest rate for a thirty-year municipal bond will typically be higher, say 5 to 7 percent. This is where rating agencies come in: Companies like Moody's or Standard & Poor's dig into the city's financial details, along with its justification for the costs, to arrive at an overall assessment of the safety of the investment. The agency then grades the bonds, like a teacher would grade a homework assignment. Safe investments are given A's while risky investments are given B's and C's. Cities that get an A rating can offer lower interest rates on their bonds because they are considered more likely to pay back investors over the lifetime of the bond; cities given a lower rating will have to pay higher interest rates to lure in investors to take on the additional risk. But here is what makes municipal bonds boring: Most of the municipal bonds issued in recent decades were rated at the highest grade—A and above—and had default rates of at most 0.03 percent, only a fraction of the default rate of private corporations. That is, if you bought a Grade A sewer bond, you had a 99.9 percent chance of getting your money back.[22]

But matters can become more complicated, as in the case of swaps. At their simplest, swaps are agreements whereby one party pays a fixed interest rate to a second party; in return, the second party pays a variable interest rate back to the first. That is, two investors exchange risks. The first party is taking on a known risk (fixed rate) while the second is guessing that over time, the variable interest will fluctuate in such a way as to be more profitable than the fixed rate. Because fixed interest rates are known, they are often more expensive—that is, have higher interest rates. For homeowners, getting a mortgage with a fixed interest rate means agreeing to a slightly higher interest rate than that offered by a mortgage with a variable interest rate; the borrower is gambling that interest rates will rise over time and so will accept a slightly higher rate at the beginning to avoid potentially much higher rates over the duration of the mortgage.

This is where swaps come in. If interest rates are particularly high when a borrower needs money, or if the borrower is sitting on a loan that already has a high interest rate, the borrower may want to do something to reduce that rate. Borrowers with a high fixed rate would be interested in a swap that exchanged their high fixed rate for a lower but variable rate. This inherently increases risk. If a city has twenty years of bonds to pay, and each year requires an interest payment of 5 percent (i.e., fixed rate), that city may be interested in swapping with a bank that would pay a variable interest rate that was starting at some lower rate (say, 4 percent) but could potentially rise above the fixed rate over time.

Just as Reagan was transforming the national economy, interest rate swaps were transforming the global economy: private sector investments using swaps rose from $3 billion in 1982 to over $6 *trillion* in 1993. During the rise of this increasingly creative financing, state governments did not allow local governments to use swaps; all work had to be financed using plain vanilla bonds. But in 1987, both California and Florida passed laws allowing governmental units to use some of these emerging financial tools to maximize their use of existing public funds. That is, to leverage via swaps.

Great savings could be had in using these finance mechanisms: in 1994 the Eastern Municipal Water District in Riverside, California, refinanced its bond debt with an interest swap—and in so doing saved the water district almost $2 million. By 2010, forty states had passed similar laws authorizing municipal authorities to engage in such exchanges, and swaps became an integral part of municipal debt markets; these transactions accounted for an estimated $250–$500 billion. These types of transactions often had their desired effect for the municipalities: they decreased the interest rates municipalities had to pay. Yet there was always a hidden risk. Fixed-rate bonds were secure, and in a sense they separated the finances of municipalities and local governments from the vagaries of the global economy. Interest rate swaps more tightly coupled the sewers with the complexities of Wall Street, and it was unclear whether many local politicians actually understood their implications.[23]

All of these dynamics were coming together simultaneously at the Cahaba River. In 1986, Jefferson County had the highest rating available for its finances: straight A's. The county's first bond issuances in 1997 and 1998, in response to the Clean Water Act requirements for sewage overflows, were for over $600 million, using fixed-rate bonds. As costs for the project began to rise, additional debt was rapidly issued for $950 million and then for another $275 million. This debt was all in forty-year, fixed-rate bonds with interest rates around 5 percent. Long-term, stable, boring, vanilla, municipal finance. Then, in 2002, things changed. The county issued its first variable-rate bonds. They did it again in 2003, for a whopping $2.24 billion. Some of this new finance was used to pay off the previous debt; that is, the fixed-rate bonds issued previously were paid for by a larger amount of variable-rate bonds.

Unfortunately for Jefferson County, the swap it had made and the bonds it used to make the swap were riskier than they realized. In the heady days of the roaring 2000s, the county had assumed it would always be profitable to be connected to global market fluctuations and decided to use an auction process—an auction rate security—to sell its bonds. Typical bond issuance is a singular event; bonds are issued, and various buyers purchase the bonds. Some buy bonds that will mature—be paid back—in five years; others buy bonds that will mature in forty years. In these vanilla bond offerings, the seller knows the interest rates for the entire debt package at the time of the selling, so the city or county knows exactly how much to budget each year for the next forty years. In the approach Jefferson County took, bonds are auctioned at regular intervals—anywhere from every week to every month—to buyers from all over the world. At a car auction, if there are many buyers, the high demand increases the price of the car. But in bond auctions, high demand means that buyers are willing to buy the debt at lower interest rates. If many buyers are interested in Jefferson County's sewer bonds, the auction may start selling the bonds at 5 percent; but as bidding goes on and there is competition for the bonds, the price goes down, perhaps closing at 3 percent. The more buyers, the lower the interest rate Jefferson County would have to pay. In the early and mid-2000s, the market for bonds was flush, so the interest rate that municipalities could get using an auction was always less than they could get using fixed-rate vanilla bonds. Jefferson County used this auction approach to convert its existing long-term debt with a high interest rate to short-term debt with a lower, variable rate of interest.

Using the auction, however, meant that Jefferson County now had significant debt in the form of variable interest rates, which made the county nervous. So, it went to its bank, J.P. Morgan, and swapped interest rates: The county paid $25 million up front to J.P. Morgan for the ability to pay the bank a fixed interest rate of 3.678 percent. In exchange, J.P. Morgan agreed to pay Jefferson County a variable interest rate pegged to a global bond index. Interest rates had been inordinately low for years, and the county naturally assumed, per J.P. Morgan's counsel, that interest rates would inevitably rise. If they did rise, J.P. Morgan's payment to the county would exceed the amount the county had to pay the bank and thus generate revenue for the county. All of this made perfect sense for the county as long as two things happened: investors kept buying bonds, and global interest rates started to rise.

What happened if there were fewer and fewer buyers? In that then unheard of worst-case scenario, the lack of sale would trigger the bonds to be sold at some very high fixed rate—perhaps as high as 15 percent. As a safeguard to prevent the auction from entirely failing or the rates from going too high, Jefferson County relied on its underwriting bank, J.P. Morgan Chase, to step in and buy the bonds at an agreed-upon maximum rate if there were no other buyers when the auction was set to close. Of course, that agreement assumed the bank had money.

By 2005, more than 25 percent of outstanding municipal debt in the United States was locked into auction rate securities and interest rate swaps. Local governments, like Jefferson County, were leveraging their infrastructure systems as capital to generate "synthetic revenue" through increasingly complex financial markets. All of this debt, however, tied local revenue and spending to global credit markets, in the latest shift in U.S. government finance. Water infrastructure finance had gone from state and local dominance in the nineteenth century to federal dominance in the twentieth and back to local just before the turn of the twenty-first century.

At the beginning of the twenty-first century, global finance played the central role. The use of credit swaps to generate synthetic revenue also signaled a significant philosophical transformation: no longer was wastewater infrastructure a public service, to be funded solely through tolls or property taxes in the region associated with that service. Now credit swaps were generating revenue, and entrepreneurial cities were using infrastructure as their collateral for entering the global finance market. During the first five years of the twenty-first century, Jefferson County had grown disturbingly entrepreneurial. The county's sewer system debt had grown to over $3 billion, and the county began using swaps even more aggressively; between 2002 and 2004 alone, the county entered seventeen swap agreements with a total value exceeding $5.8 billion. That is, the swaps were more valuable than the bonds they were being used to hedge. With almost $6 billion in swaps, Jefferson County (population 660,000) had only $1 billion less in swaps than the entire state of New York (population 19 million). Far from using swaps to hedge fluctuations in interest rate, the county was using them to speculate on those fluctuations. Its chips were all in. The county was completely dependent on global interest rates rising and auction prices falling.[24]

In January 2008, both Standard & Poor's and Moody's gave the county their highest credit rating. Only a few days later, on Valentine's Day 2008, the roof caved in. The global market froze as the housing market collapsed at the opening of the Great Recession. The debt auction went silent.

With no buyers in the auction at the expected rates of around 3–4 percent, the default base interest offering jumped to 10 percent. The new rate added $700,000 per week in interest payments alone to Jefferson County's expenses, which it was nowhere close to being able to pay. At the same time, alarmed by the global market freeze, the federal government infused enormous amounts of money into the financial system, causing the global bond index, ironically, to fall. As the index on interest rates dropped, the interest rate swap with J.P. Morgan backfired: Jefferson County had to continue paying the same fixed amount to the bank each month, but the amount that the bank had to pay the county plummeted. The last line of defense for Jefferson County was that their bonds were insured. But their insurers were also in the business of insuring home mortgages. As the subprime mortgage crisis hit, the bond insurance companies disappeared—along with any new bond buyers for the auctions. Every assumption Jefferson County had made about the financial system had been fundamentally wrong. Only eight days after the rating agencies had given Jefferson County sewer bonds straight A's, both S&P and Moody's gave the bonds junk status.

Jefferson County stumbled along for a few years before finally filing for bankruptcy in November 2011. The Securities and Exchange Commission (SEC) sued J.P. Morgan, which paid a penalty of $25 million to the SEC and $50 million back to Jefferson County, and then walked away from nearly $650 million in swap cancellation fees that the county was supposed to pay the bank. By 2011, several county employees and former contractors had been indicted by federal prosecutors in connection with the sewer project's finances for corruption and bribery. Despite all of this shady dealing between the county and Wall Street, the county had to provide the still-necessary funding for wastewater infrastructure as well as its enormous financial miscalculations; residents of Jefferson County saw their sewer rates triple, taking the average sewer bill from about $60 per month to almost $200.[25]

Water pollution is expensive. So much money is needed to deal effectively with sewage that financing the necessary infrastructure is inseparable from larger questions of how the entire government generates and spends revenue. This reality of sewers and polluted rivers shows that ideological, political battles have fiscal realities ranging from Chicago innovating the "special economic district" to FDR unleashing unheard-of sums of federal funds for treatment plants to Jefferson County using its sewer system to finance absurdly complex financial instruments. Throughout American history, paying for sewers has resulted in tectonic shifts in the political and financial structures.

Why is it that sewers are often at the cutting edge in finance? Because having clean water in a developed country is a nonnegotiable expectation. A government is not considered competent if its citizens cannot drink the publicly provided water without dying. And since the mid-1900s, a government is not considered competent if the rivers and streams under that government are little more than flowing cesspools. These expectations create a nationwide, perpetual demand for water and sewer projects, and the projects are inordinately expensive. As a nation we have not just developed new ways to keep our water clean but also, by necessity, developed new ways to finance those efforts. Thus innovation in governing rivers goes hand in hand with innovation in how we finance public investment.

PART FOUR

REGULATION

CHAPTER 9

Regulating Power

O n January 23, 1979, a group of federal bureaucrats sat around a table to decide whether a species of fish should continue to exist. The meeting started at 9:00 in the morning, and by 9:45 the group had voted unanimously to kill a dam and save the fish.[1]

This dam-killing committee was formed to help reconcile the conflicted American ideas of what the federal government's role in managing—regulating—society and the economy should be. For over two centuries the power produced by rivers via dams was the backbone of America's industries, from mills grinding flour for colonial villages to factories manufacturing weapons during World Wars I and II. The centrality of power to society required America to grapple with the relationship between private enterprise and government: How would America regulate something that society depended on?

Regulation is used by government to restrict individual rights and property for the common welfare. Every government must negotiate how to encourage economic and technological development while also ensuring public safety and a certain amount of stability and competition in the economy. Regulations, and the rights they affect, are therefore not static: as industries, technologies, and society evolve, the range of rights considered sacrosanct and those the government deems necessary to regulate also shifts. In the eighteenth century, government had to regulate the bread bakers and innkeepers; in the late twentieth century, it was telephone and pharmaceutical companies; in the twenty-first century, it is drones and the Internet.[2]

Among the aspects of the economy and society that the United States has regulated over the years, hydropower has had a disproportionate effect in shaping the government's ideas about how to approach regulation. Dams have played a central role in many societies, but their impact on American ideas of regulation has been particularly strong; the United States came into existence—and grew in geographic and economic scope—in step with the industrial revolution. Just as America was developing its political ideology, the industrial revolution was changing how power was used—in ways that once were unimaginable. And while England's industrialization was powered by coal and steam, America's ran on dams.

To appreciate how essential dams were in the nineteenth century, simply look at the 1840 U.S. Census: It found that almost every river had a dam, and many rivers had dozens. In total, the twenty-six states that made up the United States at the time had around 65,000 dams. With a population of only 17 million at that time, the United States had one dam for every 261 people.

One of the highest densities of dams in the young country was found on New England's Merrimack River, which culminated in a series of dams in Lowell, Massachusetts. Whereas cities like New Orleans were built by draining land and building on top of whatever dry patches could be scratched out, and cities like Los Angeles or Phoenix had to bring water into a city through great canals and irrigation ditches, cities on the Eastern Seaboard—like Lowell—were built on a series of engineered miniature waterfalls.

Lowell is located at the confluence of the Merrimack and Concord Rivers, just downstream of Pawtucket Falls, where the Merrimack drops through 30 feet of cataracts—the Fall Line. Timber was the predominant commodity on the upper Merrimack River, and loggers moved this commodity in vast rafts down the river. At Pawtucket Falls these rafts had to be halted, broken up and hauled by oxen over land around the falls, and then reassembled into rafts to be floated on downstream to the coast. As at most other significant falls along the rivers of the Eastern Seaboard, before long a small company built a lock and canal to get boats and rafts of timber around the falls more easily.

Even though the Fall Line was a chronic barrier to navigation, it was also a source of power. Both at the Fall Line and in the mountains upstream, rivers plunged through the sort of rapids and moderate waterfalls that were rare in England but ubiquitous in America. Heavy precipitation along the Eastern Seaboard created strong water flow that, when combined with the steep vertical drop to the coast, made the rivers ideally suited for turning waterwheels, the leading technology of the time. Colonists all along the East Coast initially put waterwheels to work in mills to process timber, which was essential for building settlements and one of the key raw materials that was plentiful in America but in short supply in Europe. By the time the earliest sawmills were built in England in the 1660s, several hundred were already being used in colonial New England. The colonists processed some timber for lumber export to England, but they used much of it for building their own settlements.[3]

Also ubiquitous around the Fall Line were gristmills, indispensable to colonists for grinding wheat to flour or corn to meal. Without a nearby mill, settlers had limited options: they could either transport their grain overland to the closest settlement with a mill for grinding, or they could grind it themselves by hand or with their limited livestock. Either option was time and labor consumptive. Grinding a single bushel of wheat into flour required about two full days of human labor, or a few hours for a horse-drawn mill. Contrast that with the typical eighteenth-century water-powered gristmill, which could grind dozens of bushels of flour or cornmeal per day. Being located near a mill offered settlers relief from some of their most demanding tasks.[4]

Mills also increased the value of exports. Along the Merrimack River, the presence of a sawmill at Pawtucket Falls in Lowell meant that wood could be exported as masts or lumber rather than as timber. At the mouth of the Merrimack, demand for masts was particularly high due to the burgeoning shipping business. Similarly, all along the East Coast, gristmills allowed communities to export flour, which was easier to transport and more valuable than wheat, at downstream ports. More than any other type of infrastructure, sawmills and gristmills were the centerpiece of the colonial economy.[5]

Because mills were so central to early U.S. society, they had to be regulated. In many ways, a mill was comparable to a modern water, sewer, or electrical system: it provided a necessary, widely used service in the community. A mill was also a natural monopoly; rural communities typically only needed one gristmill, and once it was built, no other mills would move into the area. Thus, unless the government intervened, the initial mill could charge exorbitant fees. And indeed, early colonial communities and states used regulation, in place of "the invisible hand" of the market, to provide the benefits that competition normally would bring.

The government granted each authorized miller a franchise, designating them the sole provider of the service in their community. Millers with franchises were guaranteed that no other miller could block the waterway upstream or build a dam downstream that would interfere with their operations. They were also given the right to inundate upstream landowners—to build a dam whose water would be impounded on other people's land. In return for these privileges, millers agreed to fulfill certain responsibilities. Most common was the obligation to serve the public without discrimination—the millers had to grind grain for anyone who was able to pay. They were also generally required to submit to government or community oversight to ensure some minimal standards of service. Finally, because millers had a monopoly on an essential public service, their rates were set by regulation.[6] This instance of government regulating a needed public service provided by a private industry foreshadowed the role the government eventually assumed regarding public utilities; thus, gristmills have been called America's first public utility.[7]

This was the context for the humble beginnings of the Proprietors of the Locks and Canals, a small business founded in 1792 along the Merrimack River. It began as a canal company, operating a series of locks and canals to move the constant rafts of timber coming from New Hampshire around the Pawtucket Falls and send them downriver to Newburyport, then a shipbuilding center of New England. The locks and canals precisely controlled water levels in the canal to keep the logs, rafts, and boats moving up or downstream. In 1821 a group of investors more interested in power than transportation assumed ownership of Proprietors. They began assembling the land and capital necessary to build Merrimack Manufacturing Company, a textile manufacturing company that would take advantage of the potential waterpower at the site.

Unlike many other canals, the Pawtucket Canal built by the Proprietors did not run parallel to the river. Rather, it bent away to form a half-circle meander to the south. The river and canal together created an island with the river and falls on the north and the Pawtucket Canal to the south. The effect is somewhat like a steep staircase (the waterfalls) going down to a sidewalk (the Merrimack River) at street level on one side of a park. An elevated walkway (the canal) is on the other side, with a series of elevators (the locks) that incrementally take pedestrians down to street level along the walkway so that it eventually meets the sidewalk at the far end of the park.

This geography created an unusual opportunity: if, instead of gradually lowering the water level with locks, the water level in the canal was kept elevated all the way around the island, there would be a high and constant drop in water level from the Pawtucket Canal south of the island to the lower water level of the river north of the island below the falls (i.e., the walkway would be higher than the sidewalk for the entire length of the park). By then building smaller canals that drew off portions of water from the primary, high-elevation canal, the company could create a network of waterways that, while only marginally successful as a passage for commercial traffic, was optimally designed for producing hydropower.

The Merrimack Manufacturing Company, which also owned the land and water rights of the original Proprietors of Locks and Canals, quickly found that it was managing two different sets of operations: textile manufacturing and power production. When a new manufacturing company proposed moving into town, it needed power that was owned by Merrimack Manufacturing—and Merrimack Manufacturing recognized that it had vastly more power than it alone could put to use. In 1825 the Merrimack directors decided to separate the firm's operations functionally: the remaining undeveloped land, the machine shop, and the waterpower rights were owned by one company—the Proprietors of Locks and Canals—while the Merrimack Manufacturing Company owned just one of what was

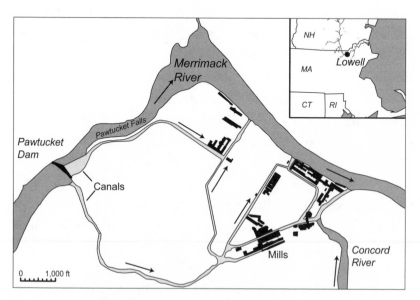

The Merrimack River and the power-generating system of canals and locks.

to become many mill complexes. In this way, the Proprietors could then supply the hydropower generated by its system directly to consumers—in this case, the manufacturers of Lowell.

This plan was a significant intellectual pivot that would establish the modus operandi for power companies of the future. Previously, manufacturers using hydropower had been obligated to provide their own power. They built their own dams, canals, and waterwheels to power the equipment, whether it was a saw to cut lumber or a spindle for the emerging industry of textile manufacturing. The Proprietors' approach was to take care of the power part of manufacturing by managing the volume and fall of water through and across the island at Lowell. With their system of dams, locks, and canals, the Proprietors could guarantee a precise amount of energy to each manufacturer in Lowell. The manufacturers, then primarily the owners of textile mills, were still responsible for constructing the waterwheels necessary to convert this potential energy into usable power. But they no longer had to construct, maintain, and operate the dam, gates, and channels. Instead, they were able to concentrate their efforts on optimizing the use of available power specifically for manufacturing while the lock and canal company focused on greater hydraulic control of the river.[8]

This arrangement was, as the historian Patrick Malone notes, one of the earliest forms of a power market—the potential energy of a river was commodified, bought, and sold in units of power rather than in units of product produced. The textile manufacturers paid the Proprietors not for a fraction of the river flow, but for a "mill-power" equal to 25 cubic feet of water per second dropping 30 vertical feet. This was about the same as 85 horsepower, enough to power a mill with over 3,500 spindles and all other machinery necessary to turn cotton into cloth. Just like modern power companies supply power—electricity—for industries or individuals to then use for their own purposes, so the Proprietors supplied manufacturers in nineteenth-century Lowell specifically with waterpower, and then the manufacturers had to put that energy to work. While the location on the Fall Line made operating a mill in Lowell attractive, it was this ability to purchase a set amount of potential energy that drove the textile industry to such heights in Lowell. Following this development, other companies adopted similar approaches—first in the Merrimack valley and then more broadly in New England—severing water use from the products being made and thus turning that water into the commodity of power.[9]

This change in the conceptualization of power was significant for industrializing the economy, but just as important was a change in the regulation of power. In 1833 the Proprietors sought to increase power production by increasing the height of the dam on the Merrimack by two feet. Raising that dam would back up water farther upstream, inundating upstream riparian landowners. In fact, the Proprietors' proposed dam raising would inundate lands all the way to Nashua, New Hampshire—itself a growing hydropower-based manufacturing center. But the Proprietors had no reason to think this would be a problem. The government regulations in place favored economic development; smaller mills, like those upstream in Nashua, simply had to make way for the larger and more economically productive downstream mills like those at Lowell.[10]

Regulation of mills in the mid-nineteenth century had taken a distinct turn away from its original mission of ensuring minimum subsistence services and toward full-scale economic development. Private property was being sacrificed for the broader economy through the emergence of regulations known as mill acts, which gave dam builders and owners what amounted to eminent domain along rivers. The legal historian Milton Horowitz viewed this shift from protecting private property to sacrificing property in the name of economic development as one of the great transformations of U.S. law.

Giving millers—and, eventually, hydropower companies like the Proprietors—rights of eminent domain required a substantial shift in the basic concept of property in America during the second quarter of the nineteenth century. This in turn promoted the view that government regulation should be based on considering the relative efficiencies of conflicting property uses. That is, the government should decide whether society, or the local community, would be better served by small dams and secure property rights in contrast to large dams and economic growth—and the accompanying insecure private property rights. Without such a paradigm change in interpreting and implementing regulation, the primacy of riparian property rights would have minimized opportunities in river development for hydropower. Instead, the desire for economic progress in the early nineteenth century was so strong that it shifted government regulation from favoring established property to supporting whichever use best served economic development. Thus, with the government's encouragement, dams and mills proliferated in number, complexity, and economic output. As a result, textile manufacturing via hydropower in America quickly surpassed that in England.[11]

But textile manufacturing, typically associated with nineteenth-century hydropower, was secondary to flour milling. As settlers cleared greater amounts of land for agriculture, grain became plentiful and gristmills were increasingly converted from low-power community mills into high-power merchant mills that produced thousands of pounds of flour per hour and then sold it on the broader Atlantic market. These sales were facilitated in part by geography; many of the larger mills were located at or near the Fall Line, which was the upstream extent of navigation for Eastern Seaboard rivers. Farmers upstream of the Fall Line brought their grain downstream, where the burgeoning mills processed it into flour, and then dock workers loaded it on the oceangoing ships waiting there at the head of navigation. Because a single mill could manufacture 150 barrels of flour a day, flour export expanded dramatically. And when consolidated into industrial facilities, the export market exploded: the fifty mills within eighteen miles of Baltimore kept ships moving through its harbor, as did other ports on the Eastern Seaboard that had similar ratios of mills to harbors.[12]

Throughout the nineteenth century, rapid construction of water-powered mills, adoption of new technologies, and a booming population's demand for a variety of products led manufacturers to expand their river-powered industries beyond just grain and wood. In 1826 Delaware County, Pennsylvania, counted 129 mills: 53 sawmills, 39 flour mills, 14 woolen factories, 12 cotton yarn factories, and 11 paper mills. On the Brandywine River near Wilmington, Delaware, E.I. DuPont constructed a mill for manufacturing gunpowder in 1802, became the largest producer of gunpowder by the mid-nineteenth century, and later used that success as a springboard for becoming a chemical manufacturing empire.[13]

These waterpower-based industries had a lasting effect on the landscape of the United States. Typically, the cities that developed around navigable rivers or ports had businesses and settlements strewn all along the waterfront with the supporting housing clustered closely behind. But to take advantage of the potential hydropower available in the narrow valleys of rivers at the Fall Line, the businesses had to be densely located immediately next to the falls, and they were stacked right on top of each other. Yet, on a larger scale, the abundance of potential waterpower sites diffused manufacturing. The Eastern Seaboard had a continuous line of steep streams and rivers that provided a consistent supply of waterpower throughout the region. For any potential entrepreneurial manufacturer, even if one spot on a river was already taken, another was likely available on a different river just a few miles to the south or north.

While shipping and exporting concentrated economic activity at a few central harbors or at the head of navigation at the Fall Line of major rivers, water-powered milling created smaller population centers scattered throughout the landscape. Because these nascent population centers were emerging in the hinterlands, away from the cities, housing was scarce for the thousands of workers needed in the factories. These circumstances led to the creation of what would become the quintessential mill village—a few water-powered factories at the town center, surrounded by a high-density residential area that manufacturers from New England to the Carolinas built to house their workers.

All of this economic growth was facilitated by government regulation. The mill acts, written to encourage economic development in the early nineteenth century, proved highly effective. But the legislatures that set up regulations for mills did not anticipate just how different the use of hydropower would be through the middle and late 1800s. State governments ran into difficulty when justifying eminent domain rights for millers, who were increasingly seeking profit rather than providing a basic public service. And so the pendulum of regulation swung back toward the rights of private property.[14]

Only a few decades later, state governments viewed as outright unconstitutional the same regulations that had been instrumental in allowing mills to develop basic settlements and communities in the early nineteenth century. In 1855 the Supreme Judicial Court of Maine found that under the more developed economic conditions of the mid-nineteenth century, the mill acts "pushed the power of eminent domain to the very verge of constitutional inhibition." The Supreme Court of Vermont in 1860 likewise suggested that the mill acts "stepped to the very verge of constitutional limit, if not beyond." When public service was no longer the sole or primary purpose, and when mills peppered the landscape, there was little justification for sustaining regulations that favored the construction of more and larger mill dams. By the 1870s, state courts throughout the United States were striking down their states' mill regulations, and waterpower was newly placed on the same footing as all other industries.[15]

The next fifty years would bring equally enormous changes in powering American society. And the government would go to extraordinary lengths to regulate the power industry, in part because power was now reaching households as well as all industries, making society dependent on the vagaries of the power industry and the emerging monopolies controlling it. Close behind all of this change was the next big technological innovation, which would transform the way power was moved through society—the grid.

CHAPTER 10

The Power of a River

The closing decades of the nineteenth century marked the beginning of the movement away from using rivers for power. Using steam to power manufacturing had tremendous advantages over hydropower. The greatest effect of steam power on the landscape was to shift the geography of industrialization, uncoupling the potential of power from the hydrology of a particular location. Water-powered mills could harness only a finite amount of power from a river; the flow of water at any site was naturally limited, and so there was a hydrologic cap on the potential size of the manufacturing complex at a site. Steam power, which could do the same work as waterpower for manufacturers, not only increased the maximum potential of power available at any site but also allowed geographic flexibility. Manufacturers using steam power were not forced to cluster around specific sites or venture away from the cities and start their own villages like those companies at Lowell had. Instead, they could move their production closer to the hubs of transportation and easily export their products from there. With the rise of steam power, cities like Chicago, Cleveland, and Detroit, along the extremely flat margins of the Great Lakes, were able to grow into much bigger centers of manufacturing and shipping than they could have if manufacturers had remained dependent on waterpower.

Yet the predominance of river power kept a strong, stubborn hold on American industrialization—much more so than in other countries, where the pivot from rivers to steam was almost immediate and complete. Well past the turn of the twentieth century, many U.S. industries depended on waterpower rather than switching to steam. They did so because of their investment in the basic infrastructure of dams and mills, and because waterpower was cheap and plentiful. When American industries installed steam engines, their initial goal was to provide reserve capacity—power for when river flows were insufficient to turn the turbines. England was rich in coal but poor in hydropower potential, and manufacturing in England switched almost immediately from water to steam power. However, rivers continued powering American industry throughout the nineteenth century and through the turn of the twentieth century.[1]

Although American industries were slow to shift from rivers to steam for power, they rapidly adopted the use of electric power, an innovation that was outright transformative. Electric power is the conversion of mechanical energy—whether generated by a waterwheel or a turbine in a steam plant—into electrical current, which is then delivered through a circuit and expended as power, whether for light, or heat, or motion. Generating power from electricity uses largely the same process: spinning turbines. But converting that spinning motion into electric current, and powering equipment using the simple movement of electrons, has two vital characteristics. First, electrons are fully generic. Electric current generated by a turbine on a dam is fully interchangeable with current generated by a windmill or a coal-fired power plant. Second, electricity can be transmitted over enormous distances—hundreds or even thousands of miles. The combination of these crucial characteristics means the lightbulb in a house is as likely to be glowing with electrons from a hydropower dam in Georgia as from a coal-fired power plant in Ohio or a wind turbine in Massachusetts. For the consumer, it doesn't matter because the result is the same. The electricity grid can thus integrate many generation sources along with tens of thousands of miles of transmission and distribution lines, creating a vast power network that eventually eliminated much of the geography of power.

Soon after Edison introduced the light bulb to the world in 1879, he developed the first electricity grid, using a coal-fired steam engine to produce the electricity to light fifty-nine homes in Manhattan. Less than a month later, the world's first hydroelectric plant, built on the Fox River in Wisconsin, generated the power to illuminate 250 light bulbs in Appleton, Wisconsin. Electricity production and use then quickly moved from novelty to standard expectation; houses were built already wired for electric lights, electricity-powered street-cars became the standard form of public transportation, and manu-facturers redesigned their factories to make use of electric motors. This shift happened extraordinarily quickly: the electrification of Chicago's factories grew from 4 percent to 78 percent in the thirty years between 1900 and 1930.[2]

Early electric companies performed all the duties involved in pow-ering society, from generation to transmission, distribution, and mar-keting (i.e., selling) the power to its end users, both industries and individuals. Because electricity is interchangeable, a single company need not perform all these functions; but by doing so, it retains con-trol over the product's entire supply chain. And in the early nine-teenth century, this meant having full control over the availability of a product that was becoming essential to society. Indeed, private electric companies were quite similar to the gristmills of the eigh-teenth and early nineteenth centuries—they provided an essential service to society and had a natural monopoly.

And, as with gristmills, the government stepped in to regulate electric companies. Three states established electric regulating agen-cies in 1907, and 33 of the 48 states had such agencies by 1916. The electric companies were given a franchise in an area (no other com-panies were allowed to enter the area) on the condition that they would provide service to anyone in that area who was willing to pay and that their rates would be set by a public utility commission to ensure a minimum level of service at a reasonable cost. Almost every rule or regulation that had been developed for gristmills of the early nineteenth century was being adapted and applied to electric companies of the early twentieth century.[3]

This type of regulation worked well for small power companies, but it was less effective when the power companies began consolidating physically and financially. In 1926 the private company Philadelphia Electric began developing the Conowingo hydroelectric project on the lower Susquehanna River, where the river crosses the Fall Line. But the potential power from this single hydropower source was so great that the electricity would dwarf Philadelphia Electric's own needs. The only way to justify building a dam that could create this much power was to increase demand. So three electric companies—Philadelphia Electric, Pennsylvania Power and Light, and New Jersey's Public Service Electric and Gas—set up a regional transmission network that allowed them to share the output of their various generating plants, including Philadelphia Electric's Conowingo. The 1927 agreement created the first integrated, centrally dispatched power pool in the nation, a vast grid spanning the entire Mid-Atlantic region that became known as the PJN Interchange. By connecting their networks, the three companies dramatically increased the number and diversity of users, and they could take advantage of distant, enormous power sources—like large dammed rivers—that produced more power than a single electric company serving a single city or region could use.[4]

The Conowingo exemplified what was happening in the industry more broadly. First, these electric companies were vertically integrated businesses in that they provided every aspect of the electricity supply chain, from generation to transmission to distribution and marketing at the level of individual homes and businesses. Second, over the first three decades of the twentieth century, individual electricity utilities were being systematically rolled into power companies that owned many utilities. These different power companies, in turn, became part of broader holding companies. Holding companies are companies that own companies; Google, for instance, is now owned by Alphabet, Inc., a holding company that owns several technology companies.

As an example, in 1925 the national-scale company Electric Bond & Share bought Northwestern Electric Company, a small regional power company in Oregon that generated electricity from the Condit Dam on the White Sturgeon River. The firm also bought Montana Power and Washington Water Power Company, which each had thirteen hydropower dams. The financial ballast behind Electric Bond & Share's purchases of these small Northwestern power companies was wealthy investor J. P. Morgan Jr. By the time Morgan's holding company was done, it controlled 53 percent of the Pacific Northwest's electricity demand—and it did not stop there; with control of 15 percent of the nation's electricity, Morgan's Electric Bond & Share was the nation's largest electric utility holding company.

As this trend toward consolidation expanded throughout the industry, the growing electricity-using population found itself being serviced by fewer and fewer companies that were owned by only a handful of holding companies. In Wisconsin, for example, 25 percent of homes that had electricity in 1917 were serviced by 312 distinct electric utility companies. By 1930 over 90 percent of homes had electricity, all supplied by a mere nine companies—eight of them owned by the big three holding companies in the region. These holding companies created numerous problems for government regulators.[5]

To progressives of the early twentieth century, private electric companies were the most recent form of a trust—like the railroads of the late nineteenth century or oil of the early twentieth—that needed to be broken up or regulated. To government regulators, private industry had gained too much control over too many central functions of society; the entire backbone of modern America was under the control of Wall Street tycoons like J. P. Morgan Jr.

Progressives like Theodore Roosevelt bristled at the fact that a handful of power companies were monopolizing rivers in addition to electricity markets. Because they were developing and growing far more quickly than public power companies such as cooperatives, private companies were able to snatch up most of the ideal dam sites, further tightening their stranglehold on the industry. Adding insult to injury, because regulated private electric utilities were required to provide service to all in their region, the private companies avoided rural regions, where potential profits were lower due to users being widely dispersed and largely residential. Cities contained high-use industries as well as densely populated residences, which minimized costs of transmission and distribution and thus maximized revenue. Therefore the growth of private electricity in the United States through the 1920s was private, hydropower-driven, and focused on powering cities to the exclusion of rural regions.[6]

The progressives sought a way to simultaneously break up these electric trusts, electrify rural areas, and reassert public use and development of the remaining prime dam sites. The solution developed by Franklin Delano Roosevelt was for the government itself to start an electric company that operated in direct competition with private industry. Thus the Tennessee Valley Authority—or TVA, as it would become known—was founded to supply the power that it generated at its own dams to a rural poverty-ridden area: the Tennessee Valley, in the Deep South.

The TVA is unusual, its status resting somewhere between a federal river management agency and a private power company. It has been this way since its inauspicious beginnings with the half-finished dam begun by the War Department at Muscle Shoals, Alabama, in 1918. The dam was intended to generate the electricity needed to produce nitrate, a key ingredient in the explosives used in World War I. The war ended not long after construction began, and the federal government was left with an unfinished dam with no official purpose.

The dam at Muscle Shoals would eventually be named Wilson

Dam. Its fate was tied up in the question of what role the federal government should have in the business of producing power. If the federal government produced electricity, it would be competing directly with private industry—something the government had been careful to avoid as recently as 1928, when the bill authorizing the Bureau of Reclamation to build the Hoover Dam had contained provisions to allow power production. Instead of operating the dam itself, the Bureau of Reclamation leased the power potential of Hoover Dam to nonfederal power companies: the bureau provided the stored water volume and the hydraulic pressure to the power companies, which handled the transmission, distribution, and marketing themselves. This arrangement kept the federal government from any semblance of competing with private or local power. The half-built dam at Muscle Shoals provided a similar opportunity to repurpose a federal dam for private power, and among the various ideas regarding what to do with the dam was an offer from Henry Ford to buy it for a proposed new automobile-manufacturing hub in Alabama.

Instead, the dam, which took the Corps of Engineers six more years to complete, remained under federal control and followed a path similar to that of the Hoover Dam: power produced from the dam by the Corps would be sold under a ten-year contract to a regional private electric utility, Alabama Power and Light. This created a public-private partnership in which power generated using a federal dam was distributed through private transmission lines. But the long-term future was uncertain. Alabama Power's contract would be up for renewal in 1935, and there was a growing desire by many in the federal government and anti-industry groups to use this dam and many others to compete with Alabama Power.[7]

These were the starting points of a years-long battle between Wendell Willkie and David Lilienthal. Willkie's push for continued growth in the private power sector clashed with Lilienthal's vision of government regulation—or even replacement—of the private sector. Willkie was the face of private power in his capacity as first an attorney for and then the president of Commonwealth & Southern Company, the

(Left) Southern utilities executive Wendell Willkie, ca. 1940—the year he was the Republican nominee for president. *(Right)* Tennessee Valley Authority Director David Lilienthal, ca. 1938.

holding company that owned Alabama Power & Light. With a full frame and a full head of dark, somewhat unruly hair, Willkie was eloquent and charming but also was described as arrogant, cocksure, and impetuous.[8] His nemesis, attorney David Lilienthal, had come to the Tennessee Valley to do battle with Commonwealth & Southern and everything it represented. Though temperamentally similar, Lilienthal was physically the opposite of Willkie: bald, trim, and elegantly dressed.

While Willkie made his mark in private practice, Lilienthal's career was driven by his tireless efforts to use government regulation to rein in private power. During his brief years in private practice immediately after graduating from Harvard Law School, Lilienthal authored a string of law review articles that delved into the weeds of regulation, along with articles for liberal and progressive magazines such as the *New Republic.* Two articles in the *Columbia Law Review,* in particular, showed Lilienthal's deep legal scholarship. He addressed the issues that states faced when trying to control the practices of holding companies that crossed their borders: states were relying on regulations designed for individual plants when what they needed were regulations designed for entire systems. Lilienthal's writings quickly drew the attention of progressives interested in limiting the growing influence of holding companies in the power industry, which typically were referred to as the power trusts.[9]

Lilienthal took his first step into government regulation when the governor of Wisconsin hired him to head the Public Services Commission. In that position, Lilienthal could implement his vision of regulating the system of industries rather than individual companies. He quickly earned a reputation as an aggressive pit bull, but one with extraordinary understanding of the intricacies of industry, regulations, and how regulated industries set their rates. This last aspect was critical: because regulated industries like water and sewer service or electricity have natural monopolies, the rates they charge are set by negotiating with regulatory commissions, which thus need expertise in rate setting along with a willingness to take on industries that constantly try to raise rates.

In one of his first and most contentious rate cases, in which he battled with Wisconsin Telephone Company, Lilienthal uncovered irregularities in the company's relationship with its holding company—AT&T. He showed that the company continued to generate profits while the public was in the midst of the Great Depression, thus posing the question of whether rates should reflect the profit needs of private industry or acknowledge the realities of the customers' ability to pay. In previous cases placing government regulators against private industry, the private sector often relied on deep expertise in their subject that the public sector typically lacked. But in this case, Lilienthal brought in a series of economists; he invited professors from the University of Wisconsin, the University of Chicago, and Princeton—one of whom was past president of the American Economic Association—to testify against the fiscal rationale adopted by Wisconsin Telephone. As in almost every aspect of his management during his time at the Public Services Commission, Lilienthal set out to challenge the status quo by questioning the basic economic foundation of the private sector's arguments and by uncovering how the private sector continued to flourish while the public sat in fiscal depression.[10]

Lilienthal's purview ranged from the telephone industry to the railroad and electricity industries. He was also unafraid of drawing in his colleagues in the legislature, a tactic that enabled him to send shocks through the private utility companies when lawmakers simultaneously strengthened the regulatory commission and introduced a constitutional amendment allowing local governments to finance their own power projects. This was a brilliant political move; the private power companies could not fight the increased regulatory authority for fear that local governments would respond by beginning their own public power projects, which would cut into private power's potential profits and ability to expand.

After convincing private power to drop its opposition to increased regulatory authority, Lilienthal dramatically increased the size of the Public Services Commission. He allowed it to hire experts to conduct in-depth investigations into private utility practices and prices. In less than a year, the commission could take credit for up to $3 million in reduced rates for a number of utilities that affected over 560,000 customers. Still not satisfied, Lilienthal relentlessly went after corporations, threatening at times to reduce allowable rates to the point where regulated companies began to worry over their financial stability. He gained a national reputation, garnering praise from influential sources ranging from Supreme Court Justice Louis Brandeis to the editorial page of the New York Times. In all, he was viewed as a national rising star for his method of "regulation with a vengeance" when taking on the power trusts.

After only two years in Wisconsin, in 1933 Lilienthal was tapped as one of the directors of the just-created TVA. This move put him in direct conflict with Willkie. One cause of their ongoing conflict was how aggressively Lilienthal came to define the TVA's originally nebulous mission. In the midst of FDR's blitz of his first hundred days in 1933, he had asked Congress to "create a Tennessee Valley Authority—a corporation clothed with the power of government, but possessed of the flexibility and initiative of private enterprise."[11] This directive was initially interpreted as envisioning a government agency intended, variously, to improve navigation, provide flood control, produce fertilizer, generate power, and provide for regional economic development.

But as Lilienthal took control, the TVA gained the greatest traction as an unusual mechanism for regulating the private power companies. At the time, utility commissions typically had no idea whether the rates that private industry lobbied for were realistic. There was little the government could do short of grinding through, case by case, state by state, as Lilienthal had done in Wisconsin. It remained difficult to know whether the rates charged for any of the different services of private power were fairly priced. As Lilienthal realized, only if the government did the same work as a private power company from start to finish could it, in FDR's words, "discover the true cost" of power. Both Roosevelt and Lilienthal took to describing the TVA as a "yardstick" for comparing private costs and rates. When Alabama Power & Light's contract for the Wilson Dam was close to running out, Lilienthal sequestered himself for weeks with the books and estimates of how much power would be consumed. And in September 1933, he announced that the TVA would charge 3 to 4 cents per kilowatt hour, compared to the 5 to 6 cents that was the going rate of private power companies.[12]

The price Lilienthal set fulfilled the two purposes that he had developed for the TVA. Because it was low compared to the average rate of utilities in the South—and nationally—it put private companies and the public on notice that the government's rates for power would be low, even artificially so; utility stocks and bonds fell in response to the announcement. The second reason Lilienthal set the rate so low was that the TVA was already producing too much power. Lilienthal hoped low rates would give the public incentive to use electricity. If use increased and the same amount of power was still generated, the cost per unit of power would decrease even more, providing an even greater contrast to private power.

All of this news was deeply problematic to private power, which saw the TVA as setting a dangerous precedent. For the next six years, Wendell Willkie would represent the private power interests in their battle against Lilienthal and the TVA. Willkie had been critical of the TVA since it was first proposed in Congress, going so far as to testify against it during the congressional hearings. His holding company, Commonwealth & Southern, had much to lose to TVA; it owned all the common stock in Tennessee Electric Power Company, Alabama Power, Georgia Power, and Mississippi Power, all of which were in the proposed service area of the TVA. As long as TVA followed a course similar to what had been used at Hoover Dam or was currently being used at Wilson Dam, the potential conflict could be resolved. But the Wilson Dam terms with Alabama Power were up for relicensing in 1934, making it the first test case for how TVA would deal with private power. Willkie proposed a sharing relationship between TVA and private power, but Lilienthal was having none of it.

Lilienthal's efforts at regulating private industries in Wisconsin had jaded him about the intentions and trustworthiness of private power, and his growing fan base of progressives gave him plenty of support for keeping the TVA ideologically pure. More pragmatically, if private companies were still in charge of transmission and distribution, the TVA yardstick would remain incomplete—the government would be unable to measure for itself a significant portion of the costs of providing electricity. More broadly, by refusing to acquiesce on any front, Lilienthal ensured that the TVA would have considerably greater leverage over power in the entire region than if private power retained any foothold.

As he went into negotiations with Willkie over the future contract for Wilson Dam power, Lilienthal knew that he had Roosevelt's unequivocal support. The TVA was an ideal model of the policies behind the New Deal—a federal agency that absorbed enormous amounts of labor and, more philosophically, applied systematic planning to a regional economic development problem. Roosevelt called the TVA the "apple of his eye" and made regular trips to the valley to see its progress.

Throughout the Great Depression, the TVA was abuzz with dam-building activity. Each dam was an additional source of power and a direct threat to existing power companies; Willkie was negotiating for private power from a position of declining strength. Wilson Dam was significant, but the dams begun by the TVA during the Depression were equally impressive and cheaper for the TVA to build than such dams would be for private power companies. When a private power company built a dam or a steam power plant, it had to recoup all those construction costs through higher rates. But when the TVA built a dam, the federal government covered some of the costs in the name of broader public interest such as navigation and flood control. This support reduced the cost of power generation and thus reduced the rates TVA needed to charge, so Lilienthal could set far lower power rates than his for-profit private power competitors could afford.

On top of lowering the rates, Lilienthal played his hand to the max by suggesting that the TVA might start building power lines directly parallel to those already built by Alabama Power & Light—a not-so-subtle hint that Willkie could either surrender by selling his company's existing lines and system to TVA or be directly competed out of business. What's more, the smaller distribution lines—the final handoff of power from generation and transmission to the individual homeowner or business—were being built through heavily subsidized Public Works Administration loans given to a growing network of public, municipal utility cooperatives (i.e., nonprofits). The emerging power grid in the Tennessee Valley eliminated the need for private power at any step. And this intentionally combative approach to regulating the price of power through direct competition was like Lilienthal's earlier approach in Wisconsin: regulation with a vengeance.

From the beginning it was unclear whether the TVA would be able to move forward as an activity of the federal government, because it stretched ideas about the types of activities the federal government could engage in. Yet a series of Supreme Court decisions gave the TVA somewhat unexpected support. In 1936, stockholders of the Alabama Power Company sued the TVA for its proposal to buy the company's property and equipment. The stockholders hired Wendell Willkie as their attorney, and he scaled up the case into a broadside attack on the TVA as exceeding the scope of congressional power. But the Supreme Court found against Willkie. In this first test of the TVA's role, the Court ruled that the proposal from the TVA for "disposing" surplus energy from Wilson Dam was constitutional.

In 1938, Alabama Power itself sued the federal government for giving grants to municipal cooperatives that would then purchase power from the TVA. Once again, the Court ruled in favor of the TVA. And then, in 1939, five companies under Wendell Willkie's holding company sued more directly against the TVA's constitutionality in a case known as *TEPCO*.[13] During the process, the TVA was ordered to freeze its power program for six months, thus giving Willkie a brief window of hope that FDR's public power company would come crashing down. But the Supreme Court once again ruled in favor of the TVA, thus frustrating Willkie's attempts to upend the agency through the courts.

Willkie realized there was little he could do to stop the federal government from going into the power business. He could not compete with the bottomless coffers of the federal treasury during the New Deal financing the TVA's infrastructure projects, nor with the Supreme Court's unexpected decisions supporting the constitutionality of the TVA. Lilienthal's entirely government-backed system created a subsidized grid from start to finish that replicated at lower costs what Willkie and private power were trying to develop themselves. The accuracy of the "yardstick" was questionable, and Willkie began saying that "TVA's yardstick is rubber from the first inch to the last," and that "The TVA touches four states but drains the nation." Nevertheless, in the aftermath of Willkie's loss in the *TEPCO* court case, he and Lilienthal finally came to an agreement. The remaining private power properties in the Tennessee Valley, like those owned by Commonwealth and Southern, were sold to the TVA. But the TVA was constrained in its potential territorial expansion; congressional approval would be required for TVA to expand. Private power had lost the regulatory battle.[14]

Then a strange thing happened: the TVA began programs to increase power use in the region rather than just provide power. As each new dam was completed, more power became available in a market that was developing a surplus of power and a dearth of buyers. Following this development, Lilienthal subtly shifted the meaning of the yardstick from being a metric for cost comparison to being a measure of the effectiveness of government-provided electricity on a grand scale. With Willkie and private industry out of business in the Tennessee Valley, there was less need for a regulation yardstick for price comparison. Now the yardstick metaphor was used to illustrate how the TVA should be a model for the rest of the country, and the model was somewhat tautological: if electricity were provided at cheap rates, then demand would be stimulated at a large scale, which would continue to drive down the per-unit cost of electricity that could be generated cheaply through centrally planned, efficient development of the hydropower sources within an entire watershed. Lilienthal argued that the rest of the United States should use this model, measuring its success against the (repurposed) regulatory yardstick of the TVA experience.[15]

Ironically, the TVA was intended to be a model for the rest of the nation but, despite its success, it was never replicated elsewhere. In 1945 Lilienthal wrote an article titled "Shall We Have More TVAs?" for the New York Times Magazine. Flush with success, he answered with a not too surprising yes. Indeed, the original plan had been to offer federal power throughout the United States. During the 1932 presidential campaign, Roosevelt had proposed beginning with four federal power projects in each corner of the United States on major rivers: the Colorado River for the Southwest, the Tennessee River for the Deep South, the Columbia River in the Northwest, and the St.

Lawrence in the Northeast. In the end, the St. Lawrence never materialized to any degree, and the Colorado and Columbia developed into "administrations" that sold or managed power produced by other agencies, such as the Corps or the Bureau of Reclamation. Only the TVA, on the Tennessee River, would manifest the federal government's grand plan of regulating private power through competition— and for good reason. Its steep topography and abundant rainfall made its many rivers ideal for hydropower.

Although New England or even the Mid-Atlantic states may have offered similarly optimal river conditions for other river authorities, cities and towns already peppered the banks of rivers, along with their existing small dams and associated mills and industries. Building a Merrimack River Authority in the model of the TVA would have required inundating towns and cities like Lowell, Massachusetts, which was founded and grew up around the water-powered mills of the river. The Tennessee Valley was far less developed and had minimal existing hydropower. Large dams could be built without large economic impacts, or at least without political pushback. Such dams were not possible in the North, where populations and industries had already settled into river valleys. Besides the other hurdles for any agency trying to replicate the TVA's success, by the time its success had been proven there was simply no longer the political appetite for another experiment like the TVA. What fit the needs of a Deep South valley in the midst of the Depression seemed highly out of place as the economy recovered and political winds changed: when asked in 1953 what he meant by the phrase "creeping socialism," President Dwight Eisenhower gave the TVA as an example. Only two decades after its creation, the TVA was already drawing political scorn.[16]

Lilienthal's modified yardstick was realized when electricity consumption rose as prices fell. But this development was not unique to the Tennessee Valley: energy consumption nationwide grew by 65 percent from 1945 to 1965 while the real price of electricity dropped by half. As the nation became increasingly electrified, power companies scrambled to keep up with demand, adding larger coal-fired power plants as well as a growing number of hydroelectric dams. Although a mention of the TVA to this day conjures images of its signature dams—Wilson, Norris, and Fontana—by 1960, within only a few years of completing construction on most of its dams, the TVA produced more power from coal-fired power plants than from the hydroelectric dams that had been used to justify its existence and that so many resources had gone into building. In 1966 the TVA began expanding into nuclear power, like many other electric utilities of that time. In the 1980s the TVA was synonymous worldwide with the planned, systematic development of entire river basins, and it dispatched engineers internationally to share their hydropower skills. Yet by that time, the distribution of power sources for the TVA was already indistinguishable from that for any other power company— hydropower made up less than 10 percent of its total generation.[17]

That the TVA was not replicated, or that it largely became just like all other power utilities, should not minimize its legacy, and that of hydropower in general, for national development. The TVA pioneered the management of an entire river basin for multiple purposes—flood control, navigation, water supply, and especially power production. Its scientists and engineers had impeccable standards of professionalism and were leaders in the development of forestry and agriculture programs, along with floodplain management, turbine technology, nuclear power production, and many other programs. They were renowned for their rapid development of fundamental and applied science, all the more impressive for their accomplishments in what had been a rural, backwater, dirt-poor corner of the country.

What's more, the TVA's effects stretched far beyond the Tennessee Valley region: Its dams were essential during World War II, when the U.S. Army made enormous requests to supply TVA-generated electricity at a peculiar facility near the town of Oak Ridge, Tennessee. The secretive site consumed such shocking quantities of TVA's hydropower that Lilienthal, who did not know what was actually being done at Oak Ridge, requested that the site be placed under TVA's (i.e., his) direction. His request was denied, and the site remained under the management of personnel from the Manhattan, New York, district office of the U.S. Army Corps of Engineers—leading it to be called, simply, the Manhattan Project.

The TVA's growth was commensurate with that of the power industry; for much of the nineteenth and twentieth centuries, power was unquestioningly prioritized over almost all other potential uses of rivers. When milldams came along, the potential of valuable riverine real estate for turning millwheels or turbines quickly took precedence over commercial navigation, which had once been favored. When towns or farms were in the way of a TVA reservoir, the residents were moved and the town drowned. The inexorable drive for power seemed to overwhelm anything in its way; regulation was written in the service of the economy and favored the power industry. But in the 1960s, a new lobby would gain political clout, and environmental concerns began, increasingly, to have the final say in how rivers were used.

The seven men who gathered in Washington, D.C., in 1979 were members of the Endangered Species Act Committee. They met to decide the intertwined fates of a dam and a fish, earning the nickname of the "God Committee" or "God Squad" for their power to decide the future of a species. And what brought the God Committee together that first time was the TVA. The agency originally designed to be the regulator was now being regulated.

The existence of the God Committee was in many ways an indication of just how much priorities in society had changed over the past two decades. Environmental concerns became increasingly valued, potentially gaining equal footing with economic development. The environmental movement ushered in a whole new era of regulation—environmental regulation. These new regulations came one after the other like a tidal wave: the National Environmental Policy Act opened the decade on January 1, 1970, and was quickly followed by the Clean Air Act in 1970 and the Clean Water Act in 1972.

Although these new regulations put new constraints on how power companies, along with all other industries, did their business, most of them were malleable in some way. The power companies could adapt by shifting here or there, or filing a report, or paying a fine. But the Endangered Species Act of 1973 (ESA) was resolutely inflexible and would prove particularly troublesome to power companies—and the TVA especially. The ESA says that no federal agency can pursue an action—like building a dam—if that action would potentially damage the habitat of an endangered species. The act was a regulatory trump card, made impossible to subvert by virtue of its sheer simplicity.

Almost immediately after the ESA passed, a strange sequence of events led it on a collision course with the TVA. The first event occurred when a previously unknown and ostensibly unremarkable species of fish was listed as endangered. In the mid-1970s, scientists had begun documenting the rarity of different species. Some species, like the bald eagle, were known to be rare and demanded protection. When scientists documented a particular species as being so rare that its future persistence was in doubt, the federal government required the species to be "listed" as "endangered"—a designation that brought out handcuffs imposed by the ESA, which constrained actions potentially affecting that species.

Only a few months after the ESA had passed, David Etnier, an ecologist at the University of Tennessee, found an unfamiliar darter—a type of small fish—in the Little Tennessee River. Darters are rather unimpressive fish. Only three inches long when fully grown, and lacking any particularly unique features, they wouldn't generally be noticeable or attractive to anyone but an ecologist. Etnier noted some differences between the fish he found and those found in other rivers, documented the fish as a new species, and named it the snail darter for its culinary preferences.[18]

You wouldn't go fishing for snail darter; in fact, you wouldn't be aware of its presence in a river unless you were looking for it very, very carefully. Unless you're on the Little Tennessee River, you shouldn't even bother looking—the Little Tennessee River is the snail darter's only known native habitat. And Etnier found it in the sole remaining undammed stretch; the TVA had put everything else under lakes. But this undammed condition of the Little Tennessee was meant to be temporary. In fact, when Etnier found the snail darter, the TVA was just finishing the Tellico Dam, which would put this last bit of river under a lake.[19]

The snail darter and the Tellico Dam are inseparable, bound together in regulatory infamy. In comparison to other TVA dams, the Tellico is unimpressive. Most of the dam is made of earth, and there is a small concrete plug at the far end of the valley. Most TVA dams are easily spotted on the horizon due to a spider web of high-voltage power lines. These lines are conspicuously missing from the Tellico, because the Tellico produces no power. It was built as supplemental storage for a larger adjacent dam on the Tennessee River, to which it is connected by a small canal. The Tellico was intended to be in the system of dams first envisioned by the TVA as part of its Progressive Era plan to harness entire rivers through careful, proactive planning. But the TVA had just not gotten around to completing the Tellico until the 1970s, so it was one of the last dams for the TVA, but a first for the ESA.[20]

The Tellico Dam project had been pushed down the TVA dam-building priority list for decades because it was projected to create relatively limited economic benefits; as proposed in the initial plan for the Tennessee Valley in the 1930s, there would be a 10 percent gain in economic benefits relative to the costs. By the 1970s, however, the TVA was finding other ways to justify its few remaining projects, and its newest approach was to justify projects based on the potential for surrounding development. As the TVA explained it, if the dam were built, in addition to enabling a marginal increase in power production, the large reservoir and available canal would be a boon to the local community. True, building the dam would require the TVA, which had been founded primarily as a way to bring rural backwaters out of poverty, to inundate a rural community with some of the last remaining fertile river-valley farms. But assuming that industrial and residential development would follow the construction of the Tellico

Dam allowed the TVA to recalculate the estimated economic benefits of the project, resulting in higher property values. This fiscal sleight of hand was enough to justify the project, if only barely, and put TVA back in the dam-building business one last time.

Then, in August 1973, the snail darter species was documented. Upon its discovery, the small fish was almost immediately listed as an endangered species under the new, largely untested ESA. With the prospect of the snail darter's only known habitat being inundated by the Tellico Dam reservoir, Article 7 of the ESA took effect. And it offered no exceptions: even though the dam was over 90 percent completed, construction would have to stop. In an ironic twist, the TVA found itself on the wrong side of regulations.

From this point on, the story commonly told of the Tellico and the darter diverges tremendously from what the journalist Dan Rather later described as the "real story." But first, the public story. Of all the strange things that happened between the dam and the darter, perhaps the strangest was the TVA's response to environmental regulation: the TVA claimed that it was immune to other federal agencies' regulations, whether they were set by the ESA in the case of the snail darter or by the Clean Air Act when it came to the TVA's coal-fired power plants, which were also under regulatory scrutiny. The world had changed around the TVA, but the agency wasn't ready to acknowledge it. The agency that had been set up to be a model of holistic river basin development was placing the demands of power not just above all other interests, but above the federal government itself. When the TVA had been challenged in the Supreme Court shortly after its conception, when Willkie and others were contesting its very existence, the TVA invariably won. But four decades later, when the TVA challenged these new environmental regulations, it inevitably lost. In June 1978 the Supreme Court said that the TVA, like other power companies and like other federal agencies, would be regulated by these new federal laws. And specifically regarding the snail darter, the Supreme Court said that, regardless of how inane the fish might seem and no matter how far along the dam construction was, the Tellico project must cease.

This decision dominates much of the rhetoric and lore of the snail darter: the idea that a tiny fish was deemed more important than a dam. This part of the story is what makes the fish and the dam as much a part of liberal environmental lore as they are of libertarian anti-regulation lore. But it ignores the second part of the story, which is more about economics and farmers than fish.

Four months after the Supreme Court decision, Tennessee Senator Howard Baker and Iowa Senator John Culver were still furious that states could lose dams and other infrastructure projects to fish. The two senators introduced an amendment to the ESA that would create the Endangered Species Committee, or the God Committee. The committee was designed to be composed of federal leaders who would potentially balance the competing interests of a species and a development project: the Secretary of Agriculture, Secretary of Army, Secretary of Interior, Administrator of the EPA, Administrator of the National Oceanic and Atmospheric Administration, Chairman of the Council of Economic Advisers, and a representative of the state in question. The God Committee was an intentionally designed loophole to get around protecting endangered species when doing so proved inconvenient.

When it convened for the first time on that January morning in 1979, the God Committee actually had two projects and two species to consider. The first was the Grayrocks project: a dam and reservoir on the Laramie River in Wyoming that would provide cooling water for a coal-fired power plant but have an impact downstream on whooping cranes in Nebraska. The God Committee went through the case, found that Grayrocks was a valuable economic development project and in the national interest, and reviewed some of the measures proposed to offset potential damage to the species. With that balance of positive economic development and mitigation of the potential negative impact to endangered species, the God Committee voted unanimously to grant the Grayrocks project an exemption from the ESA.

Then the conversation turned to the snail darter. Surprisingly, the value of the endangered fish was not prominent in the discussion or in the group's decision. The presentation of the Tellico case for the God Committee focused almost exclusively on the economic costs and benefits of the dam project, which were not particularly promising. After describing the economics of the project, the presenter displayed a few pictures of the snail darter itself, apologizing for how difficult the fish was to see. As if to emphasize the seeming insignificance of the fish, he noted the paper clip included in the photo for a sense of scale.

After the presentation, the chairman of the Council of Economic Advisors broke the awkward silence, saying, "Well, somebody has to start." And then he summed up the equally awkward reality of the Tellico project: "The interesting phenomenon is that here is a project that is 95% complete, and if one takes just the cost of finishing it against the total benefits and does it properly, it doesn't pay, which says something about the original design."[21]

The Tellico project was found to be a boondoggle, regardless of its impact on the snail darter and regardless of how much of the dam had already been constructed. The God Committee turned down the TVA's request to complete the dam, and they did so unanimously, on economic grounds alone.

Once again, the ESA had caught Congress by surprise. And Congress responded by being far more direct: getting an explicit exception to the ESA just for the Tellico Dam. Along with Congressman John Duncan, whose district included the Tellico site, Senator Baker introduced a rider to another bill as a way to sneak the Tellico Dam through. When speaking on the floor, Baker captured what was, to him, an unintentional effect of what Congress had created in the ESA: "We who voted for the Endangered Species Act with the honest intentions of protecting such glories of nature as the wolf, the eagle, and other treasures have found that extremists with wholly different motives are using this noble act for meanly obstructive ends."[22] This rider was attached to the massive water and energy bill that the beleaguered President Carter in 1979 had little choice but to sign at the decline of his presidency.

In the popular version of the story, completion of the Tellico Dam represented economic rationality winning over environmental insensibility. But a more honest telling of the story is that the TVA had overrun its course as an instrument of economic development for the Tennessee Valley. In addition to destroying the snail darter's habitat, the Tellico Dam, like all dams, put land underwater. The Tellico also inundated some of the few remaining small rural farms in that area. In its proposal for the Tellico Dam, the TVA had based its economic rationale on the assumption that the lands around the dam would be converted from poor rural farms to power-hungry industrial users. That is, the TVA gave the industry it had originally been designed to rein in priority over the people it was designed to help: the poor farmers of the Tennessee Valley. As the reservoir slowly filled with water, these farms were inundated, literally drowned out of existence. In the end, these farms were sacrificed for industrial developments that never materialized. Today, high-value lakefront suburban sprawl and retirement communities surround the reservoir.

Regulation is about picking winners and losers; it is about setting the rules intentionally in favor of a particular group or a particular activity. From the era of textile-producing milldams up through the 1970s—over 150 years—the government indicated through its use of regulation that it considered economic development to be the primary goal. All other uses of rivers were largely sacrificed to power production.

When private power companies became too dominant, the TVA used the rivers to limit their influence. Taking a river's power potential away from the private sector and handing it over to the public sector was simply a different means to the same end of putting rivers to work generating power. The TVA's drowning of farmland to build the Tellico was a logical extension of the momentum initiated in the early nineteenth century by the drowning of the upstream landowners' property along the Merrimack and other rivers. The quest for power superseded all other concerns. Whether they involved a private company along the Merrimack or a federal agency along the Tennessee, regulations were made to ensure that power was always the big winner.

But since the 1970s—really, from the Tellico project on—the priorities of society had shifted; and with these changing priorities, the intent of regulation had to change. Hydropower development projects have become rare in the United States. While good sites for new dams are harder to find, species conservation has constrained new development significantly. Indeed, environmental conservation is now seen as a legitimate goal of government regulation that takes precedence over development and even over power production. By reining in the industries, the new era of environmental regulations established a fresh perspective on who would benefit.

PART FIVE

CONSERVATION

CHAPTER 11

Channelization

For years, from the perch of a favorite barstool in sleepy Hillsborough, North Carolina, my gaze would inevitably rest on a small field just past the county courthouse. Through the field ran a small creek, ankle deep and narrow enough to jump across. More than a ditch, but too harassed by the county's lawn mowers to be a stream, for decades the little creek, "Unnamed Tributary 1," had anonymously performed its hydraulic work of conveying the runoff of small-town America.

From the same barstool, I now stare at a wall of adolescent willows and sycamores that is twenty feet tall. Through this thicket Unnamed Tributary 1 still flows, but it is now a meandering, gravelly, occasionally babbling brook. Almost nameworthy.

Setting aside the possibility of divine intervention, I think there is something creepy about the new and improved Unnamed Tributary 1. It's too perfect. It looks the way a river is supposed to look, only small. There are mini-waterfalls, mini-pools, and mini-riffles. There are mini–gravel bars, perfectly sized for a fly-fishing Ken and Barbie. This wiggling, babbling brook, placidly flowing through evenly spaced willows and unnatural cascades, is eerily symmetrical in its perfectly contoured sinuosity.

Nature doesn't deal in symmetry or neat geometry. Streams with symmetrical meanders point to human hands. In this case, it is Adam Smith's invisible hand. The ideas that Adam Smith's hand should have a green thumb, that environmentalists should "see green in green," or that there should be a green Tea Party became almost a cliché in the first decade of the twenty-first century. Stream restoration emerged as a booming sector of America's environmental economy, ironically dwarfing the headline-grabbing but largely anemic twenty-first-century carbon markets in yet another example of how rivers and their advocates have quietly shaped the American economy. Conversely, as the American economy has waxed and waned over the past centuries, how—and whether—we moved, filled, drained, and re-meandered rivers evolved alongside it. The story of how symmetrical streams became a traded commodity and how this particular stream in Hillsborough, North Carolina, took on its perfect sine-wave curves has its origins in the nineteenth-century backwaters of upstate New York.

Among anglers, the Beaverkill River is legendary. Tucked away in the upper reaches of the Catskill Mountains, the Beaverkill was remote from the nineteenth-century urban centers of Philadelphia and New York City, though close enough for the dedicated few to fish. Through the mid-nineteenth century, as trout fishing grew in popularity, urban anglers made their pilgrimages to the Beaverkill. With popularity came impacts. By the 1870s overfishing had depleted the Beaverkill, and the idea of trekking hours out of a city for increasingly absent fish was less alluring. In the polluted mill villages of New England or in the industrial wastelands of Cleveland, there were numerous likely causes for lack of fish in a stream. But in the Beaverkill, a mecca of isolation, it was harder to blame the loss of fishing on upstream pollution or dams. It seemed that the stream itself was the problem.

What makes trout so compelling for fishermen is how finicky they are. Trout like very particular conditions—water just deep enough for cover, just fast enough to sweep aquatic insects past them, and just shady enough to be cool. But trout also need variety: a bit of deep flow, a bit of shallow flow, and room to move around between the two. Unsurprisingly, trout evolved to thrive in the conditions that stream channels develop when left to their own devices.

Streams and rivers are naturally wiggly, cascading, and tortuous. The best way to see this phenomenon in action is not in a river, but in an experimental river. Imagine a fifty-foot-long aquarium, tilted a bit so that water pumped into the upper end drains fifty feet away at the "downstream" end. Fill the bottom of the aquarium with a few inches of sand and small gravel, and you have what looks like a toy river: a flume. Fluvial geomorphologists—scientists who study the shape of rivers—love flumes.

Doug Thompson is a fluvial geomorphologist at Connecticut College; he spends his summers studying real rivers and his winters playing with his flume. Along with all things geomorphic, Thompson is inordinately fascinated by where fish live—pools and riffles—and by the strange, quixotic physics of flowing water that cause pools to form in the way they do. His flume is immaculately clean, all geared up for a set of experiments in the coming academic year. After setting up the flume for a particular condition, he runs some water through it for a while, using extremely precise velocity meters to measure exactly what is going on in his tiny pools and riffles. Studying the miniature version gives him a better sense of what might cause pools or riffles to form in real rivers.

Thompson went to Colorado State University for his PhD. It's a hotbed of river science, largely because decades ago, federal agencies constructed an enormous flume facility for studying rivers—the Engineering Research Center. About the size of an airplane hangar, the lab at the research center is filled with a dozen mini-rivers for studying the intricacies of flow, sediment, erosion, and all other aspects and effects of moving water: in one corner there's a miniature dam; in the middle, a miniature Sacramento River. Some of the classic flume experiments of the twentieth century were done here, including a particular early and simple one when river scientists were just starting to really play with flumes. The Colorado State scientists partially filled a wide flume with sand, into which they cut a perfectly straight channel. When they turned on the water, visible wiggles—meanders—began developing within minutes.[1]

"Rivers meander; that's what they do," Thompson says, standing next to his flume in the basement lab. Anything that flows—rivers, blood, Gulf Stream, jet stream—will meander. You don't even need a basement lab to see this for yourself. Just cut a shallow, straight ditch in your yard and turn the hose on for a while. In only a few minutes, meanders will start to appear and then grow into the familiar sinusoidal wave of a quintessential stream.

Suppose we get down into a real, meandering river and start wading downstream. Over a hundred yards or so, we'll notice that the level of the water we're wading through rises and falls—perhaps ranging from ankle deep to waist deep. That's because depth varies over the course of a river's meanders, and in a very predictable way. At the apex of any meander is a pool where the water flow is slow-moving and deep, often over a bed of sand. Between two pools is a riffle, where the flow is shallow, swift, and gravelly. Bigger rivers, whether the Mississippi or the Columbia, will develop similar patterns at a larger scale with the pool–riffle sequence extending over miles instead of yards. Pool–riffle–pool–riffle: leave a stream or river alone and give it time, and it will inevitably find this course.[2]

Fly-fishing—or the "gentle art," as the nineteenth-century Beaverkill regulars called it—mixes a layman's understanding of these hydraulic forces with patience and an appreciation of ichthyology: find the right flow, use the right fly, and the fish should be there. Of course, this advice assumes that humans leave streams alone, which they rarely do. Instead, people have made immense and relentless efforts to straighten streams and rivers across the United States since the early eighteenth century. Meandering streams and rivers can be

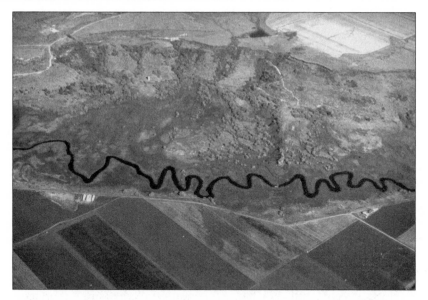

The naturally meandering Wind River in Wyoming.

sluggish and slow, and consequently they flood more often than the faster-flowing straight ones. Meandering rivers are also difficult for steamboats and flatboats to navigate. Furthermore, the driving force behind the nineteenth-century economic growth was harnessing natural resources. Nature was rarely left in any remotely natural condition.

In this push to turn the landscape from wild to productive, rivers were tweaked and turned and, most of all, straightened. Large rivers were straightened to speed along steamboat traffic, and streams throughout the Northeast and Upper Midwest—where forestry was an essential part of the economy—were cleared of gravel bars, snags, and boulders to help logs float smoothly.

Without the rhythmic pool–riffle sequence in streams and small rivers, trout-luring flow conditions would not exist, and trout would not reproduce or move to other streams. In response to the ensuing drop in the trout population, a group of late-nineteenth-century trout fishermen in the Catskills, led by the indomitable James Spencer Van Cleef, set out to fix their streams.

Like most trout fishermen in the Catskills, Van Cleef was from somewhere else. An attorney from Poughkeepsie, New York, he first visited the Beaverkill River in 1857. There he quickly fell in with the locals, who became his fishing mentors. He became a recurring presence, tromping through the Beaverkill and remnant surrounding forests on weekends and during the summer, learning the secret holes where the elusive trout still wandered and growing frustrated at finding fewer and fewer good fishing holes. As his angling prowess grew, Van Cleef became a regular contributor to the magazine *Forest and Stream*. He initially wrote about various aspects of fly-fishing, but gradually shifted toward the problem of streams and rivers. Van Cleef had a clear view of what was causing the destruction of trout streams: "I have become satisfied that the destruction of the trees bordering on these streams and the changed condition of the banks produced thereby, has resulted in the destruction of the natural harbors or hiding places of the trout." He added that this explanation for the loss of trout had the advantage of being reversible: "I believe it possible to restore most of our streams . . . especially when they are under the control of clubs or associations who can make the effort."[3]

Van Cleef had intuitively combined two important ideas that would shape—and be debated in—stream restoration for the next 150 years. First, he recognized that the physical shape of a river—its morphology—was a controlling factor for trout, and that a river's morphology was as malleable as the layout of any road. But second, he intuited the potential for private motives to restore the trout streams. Van Cleef suspected that the value of trout fishing was the experience, and he thought the trout-fishing experience was as valuable as any commodity: if people could be ensured of a good fishing experience, then they would be willing to pay some fee—an access fee—to fish that particular river. His thinking about restoration was shaped by the economic realities of the nineteenth century and his particular role as an industrial attorney. He presumed, based on the economy at the time, that private efforts were needed to restore and preserve streams, and that government would play little if any role.

In 1868 Van Cleef organized the Willowemoc Club, which began privatizing Catskills fishing areas exclusively for club members. He and his fellow club members purchased thousands of acres of trout-fishing waters in the region. The group would go on to form the Beaverkill Club in 1878 and the Balsam Lake Club in 1886, which collectively amassed more than three thousand acres, including six miles of the upper Beaverkill River.[4]

After preserving the land as private property, an approach little different from contemporary efforts by The Nature Conservancy, the clubs then turned to fixing the streams. In the 1870s Van Cleef developed rough sketches and descriptions of construction approaches, whose implementation became a hobby for club members between their fishing outings. They used what was available—logs and boulders—and placed them in particular configurations and places to create enough of the conditions for trout to be attracted to their stream reaches again. They built miniature dams—weirs—across channels to create pools upstream and fast-flowing riffles downstream. They brought in larger rocks to build veins in the river—partially submerged barriers to flow, like hydraulic speed bumps. They tried various shapes and patterns of different structures in their search to make things just right for the now elusive trout.

These were amateur efforts. Van Cleef and those who followed him in the Catskills were passionate, but they were hobbyists nonetheless. Their work was based largely on intuition and what they observed in streams they thought were in better condition. They bankrolled their own efforts and benefited personally from their relatively modest successes in actually improving trout fishing, which in turn may have increased the membership and influence of their fledgling fishing clubs. But their restoration work could never have been considered financially profitable—or even really profit seeking. Nor would they have the opportunity to test the long-term profitability of this venture—by the time Van Cleef's protégés in the Catskill fishing clubs had completed much of their work, and could theoretically have

sold club memberships for a higher price, the country was in the grip
of the Great Depression.

Wading in rivers can be a clumsy endeavor. Submerged logs, shifting sand, and swift flow can humble any novice fisherman. Rubber
waders add bulk, making a stream or river that much more challeng-
ing to navigate.

The mark of a veteran fluvial geomorphologist in the field is dex-
terity when walking in rivers. Doug Thompson, clad in chest waders,
skips and scoots down into Connecticut's Blackledge River, then up on
its banks, then down and through pools and riffles with an ease that
shows he has spent an inordinate amount of time tromping through
streams. He is as comfortable in waders as he is in his basement flume
lab. When Thompson started teaching at Connecticut College in the
late 1990s, his flume was the workhorse for his research, but he also
studied some local streams. His goal was to build an enormous data-
set of the characteristics of natural pools and riffles in New England.
The Blackledge was convenient, and so he started studying the pools
there, dragging along his students to measure all things hydraulic. The
Blackledge was where his career started, and also where it took an
unusual turn.

Standing knee deep in the Blackledge River, Thompson points out
a few of the logs half buried in the gravel of the riverbed. The logs
are largely innocuous except for one thing: they are perfectly perpen-
dicular to flow. Then he points downstream a bit toward the bank of
the river. There, a line of half-submerged rocks protrudes from the
bank into the main flow of the river, and they are slightly larger and
more angular than the rocks in the river. As we move on down the
river, Thompson points to another log, this one oriented at a perfect
45-degree angle to flow. When he first noticed these startlingly precise eccentricities a decade earlier, he realized that "something wasn't
quite right."

At the end of one of the logs, Thompson points out a rusty spike that is mostly buried in sand. Not modern rebar or a well-manufactured binding of some kind; a spike. This was the kind of spike that Thompson had literally tripped over early in his research; and after seeing the first spike, he dug around and found a lot more in logs all up and down the Blackledge. Thompson had stumbled onto the work of humans in the middle of nowhere. With a trained eye, the rest of the river comes strangely into focus: a rhythmic series of fingerlike rock piles or logs in angular structures poking out into the stream. Each of these structures was intended to deflect flow to form a small pool and riffle. Thompson had wanted to study how pools and riffles are formed naturally, but there weren't actually any natural pools and riffles in the Blackledge. This is a fairly common predicament in the heavily modified New England landscape. As he acknowledges, "It turns out working on completely natural pools in New England may not have been the best idea."

Initially flummoxed, Thompson started asking around to see who was responsible for the riverine construction work. The first answer was the local department of transportation, which had moved the stream around when building the adjacent highway in the 1950s. But the highway department had only been replicating patterns first put in place decades before. The spikes were the handiwork of the Civilian Conservation Corps (CCC).

At the turn of the twentieth century, Teddy Roosevelt and his chief of the Forest Service, Gifford Pinchot, turned the federal government's attention toward environmental conservation. The initially agrarian economy of the United States had begun growing tremendously thanks to mining and manufacturing, both of which were decimating the environment to the benefit of private industry. Roosevelt and Pinchot grew increasingly critical of this development and began envisioning ways to preserve some of the nation's natural resources.

Pinchot was a quintessential progressive; he believed societal problems could be fixed only by science, careful planning, and professionals doing the work. Stream restoration efforts by earlier groups like Van Cleef's lacked all these things; they embodied the quaint inadequacies of conservation left to amateurs: piecemeal, untested, and transient. Most of all, the approach was unscientific. That this pre–Progressive Era restoration was also being done by a group of vacationing industrialists on private club property only added insult to injury for Pinchot. He saw efforts like those of the Willowemoc Club and other amateur conservationists as inappropriate to the scale of the problem, which called for federal government leadership. Roosevelt and Pinchot's initiatives would create sprawling new mega-agencies like the Forest Service, the Bureau of Reclamation, and the Federal Power Commission, all of them focused on using the more conservation-minded potential of government in opposing the seemingly overwhelming force of markets for natural resource extraction.

The catalyzing opportunity to make stream restoration progressive, as opposed to an elite hobby, arose in Michigan when the so-called Michigan School began its work. Formally known as the Institute for Fisheries Research at the University of Michigan, this unit was formed under the biologist Carl Hubbs in 1930. Hubbs was a transplant to Michigan from California. As a Stanford graduate, he began his burgeoning scientific career studying the fishes of the Los Angeles River—which were, unsurprisingly, in steep decline. Hubbs moved to the Midwest in 1917 to be curator at the Field Museum in Chicago and then in 1920 moved to Michigan, where he was again a curator but also took the opportunity to complete a PhD in 1927. Hubbs served as director of the Institute for Fisheries Research from 1930 to 1935 and was in the vanguard of stream fisheries scientists. He ended his career back in California as director of the Scripps Institute of Oceanography, where he would be inducted into the elite National Academy of Sciences.[5]

In Michigan in the early 1930s, Hubbs had begun assembling an army of students who turned their attention to trout. For years, the decline of trout had been countered by stocking—rearing them in hatcheries and then dumping them by the hundreds or thousands into streams. This approach effectively accepted that trout could not survive to reproduce in nature in its human-modified form. Hubbs wanted to rely less on stocking and more on improving habitat conditions so that trout could naturally reproduce. In 1932 Hubbs and his students put together some of the earliest formalized, scientifically based guidelines for restoring streams. They were published in a nondescript pamphlet titled *Methods for the Improvement of Michigan Trout Streams.* This little pamphlet eventually became a cornerstone document that placed habitat improvement—restoration—on an equal footing with fish hatcheries and stocking programs for trout stream management across the United States.[6]

What Hubbs and the Michigan School actually did—what they built to restore the landscape—was largely the same as what Van Cleef had done in the Beaverkill half a century earlier. Most important, they began by constraining their efforts to working within the existing stream channel: they thought the overall shape of the channel was a set feature of the landscape, but they considered the details of the streambed as malleable—able to be designed for specific purposes. From that conceptual starting point, they constructed small dams, weirs, boulders, and logs in particular shapes and spacing. Like Van Cleef and his compatriots, Hubbs and his students were attempting to create hydraulic habitat within the confines of the existing river. But what they did next was entirely different from the haphazard work of the Beaverkill crowd, for Hubbs set out to meticulously document what happened in the streams after the structures were placed. This approach was classic Hubbs; he was a notoriously thorough and careful scientist, and for himself and his students, he intended to handle restoring streams as methodically as any scientific experiment.

The Michigan School conducted many such studies over the following two decades, documenting the changes in number, size, and feeding habits of the trout in restored streams and unrestored streams. Over the second quarter of the twentieth century, this group published dozens of articles in the primary fisheries science journals. These studies were the basis for development of longer-term evaluations and guidelines, and as students matriculated from the Michigan School and populated government agency positions elsewhere, their ideas about the utility of trout stream restoration as a technology or management technique spread nationally as well. Trout stream managers who had picked up the stream restoration bug as far away as North Carolina's Pisgah National Forest described the process as having an experimental period, when principles were developed and put into practice, followed by a period in which those principles were adapted and "modified somewhat as a result of additional information and the stress of changing conditions." The entire restoration process was conducted with a classic scientific sense of trial and error.[7]

The irony of the Michigan School's restoration efforts was that, despite all their monitoring and reevaluating and bringing in the best and brightest graduate students, few actual improvements to in-stream trout habitats were made during the mid-twentieth century; an observer would be hard-pressed to distinguish between a stream restored by Hubbs and a stream restored by Van Cleef half a century earlier. It would also be hard to distinguish between a restored stream and an unrestored stream by looking at the number of trout in each stream. If Hubbs's meticulous research and technical savvy had shown anything, it was that restoration as it was practiced up to the mid-twentieth century did not actually work.

This finding is what most fascinates—and continues to frustrate—Doug Thompson, whose office at Connecticut College is lined with books on turbulence and fluid mechanics as well as file cabinets stuffed with old surveys, maps, and data from a century earlier. Thompson was a well-known rising star in fluvial geomorphology and fluid mechanics circles when he stumbled onto that spike in Black-ledge River. After that discovery, he continued publishing papers on turbulence and river processes, but he also started spending more time in the archives of transportation departments and fishing clubs. There he discovered some of Hubbs's obscure reports, along with bits and pieces of associated data. He immersed himself in the archives of the Michigan School and other decades-old datasets that had been accumulated over the years from far-flung restored streams across the United States. He then did what obsessive scientists do when they get fixated on a particular problem: he reanalyzed the data. In 2006 Thompson published the results of this eco-historical study in a deceptively understated scientific article that concluded, "My statistical results show that the traditional use of in-stream structures for channel restoration design does not ensure demonstrable benefits for fish communities, and their ability to increase fish populations should not be presumed."[8]

Much of the data that Thompson used came from Hubbs, even though Hubbs didn't know at the time that his techniques were not working—or at least he didn't know it for certain. In later publications and reports, Hubbs started to hint at his suspicion that not all of the restoration work had been effective at increasing trout populations; but he and his students remained in a river-fixing frenzy. Why? Because stream restoration was filling a desperate need of society: it absorbed labor.

When the Great Depression hit, the effectiveness of stream restoration was unproven, but the work was perfectly positioned economically, politically, and pragmatically. Just as building levees, dams, and harbors absorbed labor on large rivers, stream restoration absorbed labor on a smaller local scale. Small stream restoration was ideal for the CCC because it required very few technical skills, was done in rural backwaters of the United States, and used local materials like logs and boulders. It was technically primitive because stream restoration as practiced by the Michigan School was in reality a re-creation of the types of structures first thought up by Van Cleef a half century earlier. The Michigan School enabled stream restoration to be included in the CCC program by promoting and justifying the work through studies published in the peer-reviewed pages of fisheries journals. The scientific publications gave these decades-old practices some sense of slide-rule perfection, which provided the grist for technical how-to manuals and pamphlets printed on cream-colored university stationery; stream restoration was thus granted academic credibility with or without any evidence of its success.[9]

As methods developed by the Michigan School were eventually adopted by the CCC and the U.S. Forest Service, the Hubbs team used workshops to teach their methods. Hubbs's assistants moved throughout the United States, leading labor crews in various regions in restoring streams. With CCC backing, stream restoration rapidly expanded well beyond its earlier confines of the Catskills and Michigan, moving into the backwaters of North Carolina, California, and Wyoming. Between 1933 and 1935, over 31,000 structures were constructed on over 400 mountain streams, and by 1936 the CCC had improved almost 5,000 miles of stream. This entire endeavor was a combination of the old and the progressive: the same techniques as Van Cleef, but put to work for Progressive Era public good.[10]

The shift from Van Cleef and his private restoration clubs to Hubbs and the Michigan School's public restoration projects was a major ideological change for stream conservation; it indicated that progressives had won an important battle in the national economy. For Van Cleef and the industrialists of the late nineteenth century, private property and limited access were essential to recovering trout; preservation required privatization. For progressives, making trout streams available to the public was part and parcel of conservation. Or, as some Forest Service stream restorers captured it, "The goal established has been kept ever in sight: The most trout fishing for the most trout fishermen, and an equal opportunity for every angler." There was no actual economic demand for restoration itself, or even for an increase in the trout population. The true results were an incremental increase in the perceived fishing experience of the general public and increased nationwide employment. In fact, the public fishing easements purchased along selected trout streams of the Catskills were intended primarily to provide the Conservation Department with another opportunity to do stream improvement work. Although the actual work being done on streams was not changing too much, the people doing and funding the restoration efforts, along with the places where they worked, were profoundly shaped by the broader forces of the national economy.[11]

That all of this stream restoration in the early twentieth century didn't work, as Doug Thompson would show decades later, made little difference to the work or perceived success of Hubbs. Even though Hubbs never analyzed the data systematically made a summary finding or recommendation, Thompson remarks, "The best indication about what Hubbs thought about restoration success is that once he got to Scripps in California, you don't hear a damn thing about stream restoration."

All of this restoration work, from Van Cleef to Hubbs, was "in-channel" restoration. The river channel itself was taken as a fixed template within which work was done. Large numbers of stone weirs and log dams might be built, but the channel itself was rarely if ever considered part of the project. But in the middle of the twentieth century, there was a growing desire to restore larger rivers; that urge started to raise the potential of working on the river shape itself, and potentially on large rivers.

Mark Twain opened his memoir *Life on the Mississippi* by observing that the Mississippi is the "crookedest river in the world, since in one part of its journey it uses up one thousand three hundred miles to cover up the same ground that the crow would fly over in six hundred and seventy-five." Meanders like those seen in the Mississippi, with their varied flows, depths, and sediments, are considered by ecologists to be the root of the extremely high biodiversity of rivers. Physical diversity begets biological diversity.[12]

Yet there are few reasons for industrial society to tolerate meandering rivers. After decades of being content with removing occasional sandbars or accumulations of logs blocking flow, river engineers of the early twentieth century set their eyes on using the increased scale of dredging and bulldozing technology available to correct the inefficiency of meanders, just as industrial foresters had removed meanders on smaller streams and rivers of New England. While dredges and bulldozers were straightening rivers, they also deepened and narrowed them to concentrate flow into as narrow a corridor as possible. On top of all of this, they removed any debris that was in the channels: logs, roots, trees, and so on. All of these activities together—straightening, deepening, narrowing, clearing—collectively fell under the ominous term *channelization*: the conversion of wide, shallow, wandering, snag-filled rivers into straight, deep, narrow, cleaned-out gutters.

Along the Lower Mississippi River, channelization was most dramatic between 1932 and 1955, when the Corps of Engineers executed its new flood control mandate by shortening the Mississippi by 150 miles. Beyond the Mississippi, between the passage of the 1936 Flood Control Act and the Clean Water Act in 1972, over 11,000 miles of rivers were channelized as part of Corps of Engineers projects. Meanwhile the Soil Conservation Service channelized over 21,400 miles on its own, often in smaller headwater rivers. One nationwide estimate put total stream and river modification at 200,000 miles, or over 7 percent of the total stream miles in the United States. In states with lowland streams and rivers more prone to flooding, channelization was even more intense: over 26 percent, or 3,123 miles, of streams and rivers in Illinois were channelized in the mid-twentieth century. And the work wasn't done when the bulldozers left: all of these efforts had to be maintained because the natural tendency of rivers to meander required constantly re-dredging the sandbars that were deposited and protecting the riverbanks that eroded as rivers sought their natural wiggly state.[13]

But for society, all of this initial work and chronic maintenance was worthwhile because channelization worked. It reduced the severity and duration of floods locally, it facilitated navigation, and it increased available land for farming. Perhaps more subtly, channelization was a river management practice that played to the strengths of engineers. Rivers could be taken from complex, unruly tangles of swamps and floodplains and converted into straight, linear, trapezoidal forms. Channelization made rivers rational.

These benefits, however, came with enormous impacts and losses to ecosystems. From destroying fish habitats to eroding banks, channelization inflicted ecological havoc. Particularly alarming was the seeming permanence of the channelization. It was widely recognized that pollution damaged the ecology of rivers, but rivers were seen as "self-cleansing": given enough time, or given enough distance downstream, they would break down their pollution. Channelization, however, was a more permanent harm. As noted at a fisheries symposium dedicated to addressing the ills of channelization:

> When a stream is polluted, its ecology is greatly changed. Like a man with a serious acute or chronic illness, its activities and functions are altered, often drastically, but there is always the hope of recovery. When a stream is channelized, it is permanently disabled.[14]

As ecologists began focusing on channelization, they were able to quantify its staggering effects. Along the Obion–Forked Deer river system outside Memphis, Tennessee, the channelization of over 241 miles was estimated to have reduced aquatic habitat by 95 percent and waterfowl hunting by 86 percent. An ecological study of a river in Iowa, where channelization reduced the length of the river from 63 miles to 34 miles, concluded with a pithy remark typical of natural scientists of the mid-twentieth century: "Results seem to clearly indicate that conditions in the channelized section were not favorable for stable populations of larger game fishes." And the impacts of such channelization on ecosystems did seem to be permanent. Fish communities had been decimated in rivers channelized over seventy years earlier from North Carolina to Missouri to Idaho, and the mammals and waterfowl that previously used the rivers and floodplains had not returned.[15]

Channelization was also not particularly cheap. Since the close of the Great Depression, the federal government had taken on a larger and larger portion of river work. Channelization was often funded by the federal government under the auspices of a range of agencies and improvement programs. As criticisms of channelization began to increase, Congress grew more attentive, leading to a 1973 congressional hearing titled "What Federally Funded Draglines Do to Our Nation's Streams." During these hearings, the Assistant Secretary of the Interior for Fish, Wildlife, and Parks stated, "Stream channel alteration under the banner of channel 'improvement' for navigation, flood reduction, and agricultural drainage is undoubtedly one of the more, if not the most, destructive water development or management practices from the viewpoint of renewable natural resources."[16]

The damage and destruction of channelization was increasingly seen as excessive; the appeal of the benefits associated with channelization began to wane when the environmental costs were factored in. A series of articles on channelization published in the 1970s included such titles as "Our Ruined Rivers," "The Stream That Used to Be," "Channelization: Short Cut to Nowhere," and "How to Kill a River by Improving It."[17]

And so the pendulum swung back, and the movement to fix channelized streams and large rivers began. At first, the mindset was to reduce the impacts of channelization by modifying the construction practices. The Federal Highway Administration recommended minimal reduction in channel length and replanting of vegetation. It also recommended replacing or supplementing gravel and large boulders in rivers that had already been channelized to approximate what might have existed before channelization. It was presumed that channelization would continue to occur, but that it could be done in a less harmful way; impacts could be mitigated.[18]

There was also growing interest in using the construction activities themselves to build habitat as part of the channelization. Channelization would continue to occur, but the hope was that inclusion of habitat structures could remedy many of its associated problems. By the 1980s, the Corps of Engineers was itself experimenting with modifying its engineered river control structures in efforts to improve fish habitats along the channelized Mississippi and Missouri Rivers.

The Corps looked for ways to keep the benefits of channelization and river engineering while also restoring some elements of fish habitat. One method involved notching the dikes along rivers—cutting gaps into river-training infrastructure. Dikes forced flow into uniform channels, so by strategically notching dikes, the Corps hoped to gain some fish habitat while retaining the overall functionality of the river engineering for barge traffic and flood control. Between 1974 and 1980 more than 1,300 notches were cut into dikes along the channelized Missouri River alone, and thousands more were cut into dikes on the Mississippi, Iowa, and Illinois Rivers.[19]

In the smaller channelized tributaries upstream, habitat structures likewise began to be built, many of them almost identical to those recommended by Van Cleef and the Michigan School decades earlier. In common with large downstream rivers, and with the streams and rivers restored by Van Cleef and Hubbs decades earlier, the restorative efforts were made via structures down in the bed of the river, while the river channel was left alone; the straightened, channelized morphology was still assumed to be a fixed characteristic of modern rivers.

For conservationists, however, these efforts worked as well as a Band-Aid would in treating cardiac arrest. What rivers needed was major surgery: if straightened and dredged rivers were ecologically moribund, but natural meandering rivers were ecologically robust, then re-meandering should reverse the ecological damage.

Re-meandering a river required much more expensive tools and more specialized expertise than Van Cleef's projects on the Beaverkill or the CCC's efforts to keep unskilled laborers at work during the Depression. The *Stream Conservation Handbook* in 1974 had recommended that "simple hand tools, a lot of muscle, sweat and ingenuity are all that is required to make a stream-improvement project" and that "stream improvement may be accomplished individually or by seeking help from Boy and Girl Scout troops and 4-H clubs." And, in a real callback to Van Cleef and the roots of restoration in fishing clubs, the author asked, "How long does it take a couple of fishing buddies to roll a likely looking boulder into midstream to form a fish shelter?" Doing work at the scale of the Lower Missouri River, the Sacramento River, or the Kissimmee River in Florida would require more than a Boy Scout troop or a fishing buddy. The movement from small-scale construction of trout stream habitat toward larger-scale river construction work would involve backhoes, bulldozers, and dredges. The trial-and-error approach used in the past had to evolve into a more engineered approach.[20]

Engineers confront increasingly complex projects with more in-depth analysis and planning. Bulldozers have to be preceded by designs, cost estimates, blueprints, construction schedules, and spreadsheets. Most importantly, for the designs, engineers need equations. Engineering is always preceded in some way by science. There are gaps between science and engineering, both ideological and temporal. The ideological gap is that engineers want knowns, specificity, and as much precision as can be wrung out of the available information. Scientists thrive on unknowns, on observing and thinking about phenomena and processes that have yet to be understood. The temporal gap exists because basic science has to precede engineering; observations must be made, systems identified, and uncertainties decreased until science is distilled to its essence: an equation. From an equation, the power of engineering can be unleashed. In the development of gasoline for automobiles, thermodynamics and chemistry had to precede chemical engineering. To reengineer a river required scientific developments in fluvial geomorphology. What was needed was not an intuitive conception of natural rivers, but rather a set of master equations for natural rivers. And in the 1950s, fluvial geomorphologists were in the midst of developing these equations.

Throughout most of the twentieth century, fluvial geomorphology was confined to the dusty pages of esoteric journals and offices of ivory towers. Geomorphology was a discipline without a home, sitting awkwardly between geology and geography, and geomorphologists were part scientists, part map readers, part scenic narrators. They were what Michael Ondaatje would describe in *The English Patient*—a book about desert geomorphologists—as those who "walked under the millimeter of haze just above the inked fibers of a map, that pure zone between distances and legend, between nature and storyteller." Quite simply, they interpreted the landscape.[21]

During the 1950s, geomorphology gained scientific legitimacy when it underwent a quantitative revolution led by Luna Leopold. Leopold was in the vanguard of a new breed of geomorphologists. He initially trained as a civil engineer and then went on to earn a PhD in geology at Harvard in 1950. Rather than mapping what he saw—as a geologist would—or describing what he saw—as a traditional geomorphologist would—Leopold's eclectic training led him to look at rivers through algebra and calculus. A particle of sediment

Professor Luna Leopold, who revolutionized river science by applying equations to rivers, ca. 1978.

was not only part of a stratigraphic layer built over indescribably long periods of time; it was a ball on an inclined plane. Thus Leopold could apply equations to the sediment particle and predict when and how it would move.

Leopold's reputation grew rapidly in the sciences, and by age 41 he was chief hydrologist of the U.S. Geological Survey (USGS). While at the agency, he worked with engineers and physicists to learn new ways to analyze data and think about the natural world. He quickly surrounded himself with brilliant young colleagues who spent the 1950s and 1960s tromping around the United States collecting enormous quantities of field data, which they analyzed according to the new quantitative paradigm that Leopold had pioneered. Professional reports and academic journals in geology and hydraulic engineering were soon peppered with articles on "quantitative geomorphology." Leopold himself analyzed the volumes of data from rivers that had been collected over the past decades, putting a quantitative spin on this remarkably unplowed intellectual terrain. And as this new breed of geomorphologists measured, graphed, and calculated their way through U.S. waterways, Leopold set out to literally rewrite geomorphology—to pen a book on the subject that would change the way the world thought about rivers.

Leopold had two co-conspirators in his work, the first being John Miller, a geology professor at Harvard who had spent considerable time with Leopold doing fieldwork in the Rockies and shared his intellectual trajectory. Tragically, Miller contracted bubonic plague while measuring soil erosion in New Mexico and died only a few days after returning to his home in Cambridge. With the book only half done, Leopold pushed forward with another author: Gordon Wolman of Johns Hopkins, known to everyone as "Reds" for his curly, wispy red hair. Wolman benefited from the same Harvard training as Leopold, but while Leopold had spent much time in the Midwest and the Rocky Mountains, Wolman was a self-described "Baltimore man," intimately familiar with the eastern rivers and the dense human footprint of the twentieth-century Atlantic Seaboard. Most importantly, he shared Leopold's penchant for the quantitative, and the two of them became fluvial geomorphology's scribblers in chief.

The combination of Leopold and Wolman was historic not just for what it contributed to the future of fluvial geomorphology, but for how it embodied the past. Leopold was the son of Aldo Leopold, author of *A Sand County Almanac* (1949), a book that elegantly mixed natural history and philosophy. In much the same way that Rachel Carson's *Silent Spring* embodied the antipollution movement, *Sand County Almanac* became the bible for the wilderness and conservation movement. *Sand County Almanac* was unfinished when Aldo Leopold died, and it was Luna who did the yeoman's work of completing, editing, and publishing the book.

All the Leopold siblings spent their childhood living in the famous "shack" in rural Sauk County, Wisconsin, the holy temple of conservation. Aldo Leopold's five children would rank highly in the who's who of environmental science for the second half of the twentieth century. Four of the five received PhDs, and three became members of the elite National Academy of Sciences. Luna was not only admitted to the National Academy but also awarded the Presidential Medal of Science for his work in rivers. Luna's intellectual interior, however, was concealed by the gruff exterior of an environmental activist—a man who would leave the Wisconsin shack for a Wyoming ranch retreat where he studied rivers and wrote. After decades with the USGS, Leopold became an infamous presence in the halls of UC Berkeley, where he terrified graduate students with his pointed questions and intolerance for sloppy thinking.

Reds was the yin to Luna's yang in almost every way. Luna's Stetson and scotch were countered by the ever-present bowtie and martini preferred by Reds. While Luna spent his formative years in a shack in rural Wisconsin, Reds spent his in the posh neighborhoods of Baltimore—where his father, Johns Hopkins professor Abel Wolman, founded the discipline of sanitation engineering and wastewater engineering. The City of Baltimore renamed its public works building after him, and the highest award given by professional engineering societies for work in water treatment remains the Abel Wolman Award. After a stint working for Leopold at the USGS, Reds returned to the same Hopkins hallways as his father. From this rarefied academic perch, Wolman also gained prominence in science—he was also inducted into the National Academy—as well as produced an entire generation of academic progeny, dozens of fluvial geomorphology PhDs who became the big names of river science in the late twentieth century and revered Reds as a gentle sage at Hopkins. Where Luna was a tall, broad-shouldered, imposing western cowboy scientist, Reds was shorter, slight, genial, gentle, and at ease with anyone and everyone.

But when they worked together, these men produced a book that was as transformative for river science as their fathers' works were for conservation and engineering. Indeed, their book, *Fluvial Processes in Geomorphology*, was an intellectual continuation of their respective fathers' work—a painstaking concoction of natural science derived from Aldo and engineering derived from Abel. This book, along with the body of work being produced by Luna, Reds, and their academic cohort, took geomorphology from narrative to equation, from descriptive to predictive, from the realm of scientists to that of engineers.[22]

The power of this new way of thinking was that it converted natural rivers into a series of equations showing that rivers were somewhat predictable. River forms could be boiled down to a series of lines on graphs. With data gleaned from seemingly innumerable studies of rivers throughout the United States, Europe, and even India, the pool–riffle sequence went from a desirable feature to a scientific rule. In fact, Leopold and Wolman noted that their volumes of data suggested that an entire pool–riffle sequence—or, as they perceived it, a complete sine-wave meander bend—would occur about every six channel widths. That is, if a river was about 20 feet wide, then every 120 feet a person walking down the river should expect to wade

Professor Gordon "Reds" Wolman, Leopold's well-matched coauthor, ca. 1972.

through a complete pool–riffle sequence, in a mathematical and predictable rhythm.

Their book was riddled with these kinds of quantifications. How tight were meander bends? How often do rivers tend to flood? How wide and deep should we expect a river to be? Human intuition was replaced with the seeds of engineering certainty. And because Leopold and Wolman were drawing on vast datasets, they gave their readers a tool with seemingly universal applicability.

Armed with this book, the river restoration movement adapted its methods. Instead of using the needs of specific types of fish— like trout—to design specific structures, the proponents of restoration began using the science of fluvial geomorphology to simulate natural conditions when they designed rivers. Channelized streams were de-channelized and then re-meandered to resemble some semi-natural version of their former selves, and the equations of Leopold and Wolman guided the design. Restoration was no longer a simple weekend hobby for fishing buddies—it was becoming an industry.

CHAPTER 12

The Restoration Economy

Interstates I-90 and I-15 cross unceremoniously in Montana. West of Butte and South of Missoula, the landscape can appear lunar on a hot summer day, the dusty open spaces punctuated only by the cars with liberal interpretations of Montana's daytime speed limit that howl past. The town—really a truck stop—at the crossing of the highways is Silver Bow, named for the adjacent creek. From the air, Silver Bow Creek stands out against the landscape: a sinusoidal wave that wanders across dozens and dozens of miles, lined on both sides by ribbons of green amid the surrounding brown. It is a re-meandered stream in the middle of nowhere.

Greg Koonce was part of the impetus behind redefining the first few miles of Silver Bow Creek, in addition to 750 other completely rebuilt rivers throughout the western United States. In the world of river restoration, and especially river re-meandering, Koonce is a relative old-timer. His company was one of the first consulting firms dedicated solely to fixing streams. In the 1980s Koonce was an early example of a new breed of consultants who served a new breed of clients in the mega-ranches of Montana, Idaho, and Oregon. Declining land prices had created opportunities for hobby ranches and vacation homes, and many subsequent newcomers to the West were avid fly-fishers.

The traditional development practice in the Rockies at the time was to put up as many houses as possible and make profit in sheer numbers. But a few developers had a different business model—they wanted to mix environmental conservation with capitalism. Koonce recalled his first interactions with these new developers: "The idea was that no one was going to restore these lands unless there was an economic reason. These new developers, however, would restore the land and the streams, and then charge more money per parcel." The idea that private property was in need of improvement was something new. Land wasn't polluted, it was just beaten up: vegetation and soil were degraded by years of overgrazing and neglect. This condition limited the potential for ranching and really anything else.

When Koonce co-founded Inter-Fluve with Bob O'Brien, Dale Miller, and Joe Urbani, they were just a group of four river scientists willing to try restoring streams. The young company rebooted the restoration industry at a substantially greater scale. Its scientists were in the business of using geomorphology to increase the chances of catching fish. Inter-Fluve advertisements in the early 1990s issues of *Rod & Reel*, *Trout*, and *Field & Stream* magazines are an almost humorous mixture of ichthyology and *Mad Men*: a picture of a trout jumping out of a stream with the caption, "We can build one for you!"

In many ways, Inter-Fluve and its clients were the modern heirs of the Catskills fishing clubs. The restoration would all be done on private land, the motivation was all about catching more trout, and it was bankrolled by wealthy landowners, many of whom were vacationers. Koonce and the others at Inter-Fluve knew about using in-channel weirs and building structures in streams for habitat. But in a significant way, their work was quite different from Van Cleef's. All the founders of Inter-Fluve went through school in the years after Leopold and Wolman's book was published. They were now armed with equations, and they could go well beyond building habitats and planting trees. As Koonce noted, "When we started Inter-Fluve, we just started building pools and riffles into existing streams with a backhoe." When these in-stream structures and built pools and riffles proved ineffective, just as they had for Van Cleef and Hubbs, Koonce wanted to try re-meandering an entire stream.

Changing the entire river was an enormous step. Koonce recalled: "The first re-meander we did was totally scary." The impetus to finally try re-meandering was part gumption and part geometry. On private land, with an aggressive developer who wanted great fishing, the scientists from Inter-Fluve had some leeway to try novel things. They were further bolstered by the reality that fly-fishing is less about the fish than it is about the aesthetics of the experience. Koonce knew that meanders were more than just natural; they were also elegantly geometric. The desire to fly-fish, and to pay to rebuild the entire river, was really a desire to improve the holistic angling experience, which meanders would enhance across the board. "The experience of standing on a gravel bar in a couple feet of water, casting into the head of a pool, and then the beautiful arc of following the pool downstream with your line; that's the essence of fly fishing," said Koonce. Fly-fishing has a natural, aesthetic use of geometry that only meandering streams can provide. And thanks to Leopold and Wolman's book, the Inter-Fluve team had the equations to size, scale, and design their meanders.

With the first re-meander under their belt, and with results proving surprisingly effective in terms of actual fish caught in the restored streams, Inter-Fluve was officially in the business of playing god with the shape of rivers. First in Idaho and then in Montana, Oregon, and throughout the Northwest, Koonce and his partners developed an entire business of sketching blueprints of rivers based on equations derived from Leopold and Wolman. Then they headed out with the construction crews to start moving rivers around—with a certain amount of horror the whole time that they didn't really know enough to do what they were doing. When word got back to Johns Hopkins and Berkeley, "Reds was flabbergasted by what we were doing," recalled Koonce. As was Luna.

For many years, Inter-Fluve took on small streams for private clients. Silver Bow Creek was one of the firm's more ambitious projects. For decades the creek had drained out of Berkeley Pit, which at the time was the largest open-pit copper mine in the world. Throughout those decades, copper tailings had accumulated in the sediment, making the river toxic and inert. Inter-Fluve led the restoration of the stream for the EPA's Superfund, a federal program that mandates cleanup of the nation's most contaminated land and ecosystems. This project was far more complicated than fixing a stream for trout. Inter-Fluve had to move the entire river to the edge of the valley, dig out the contaminated soil, replace it with imported clean soil, and then rebuild the river in the middle of the valley. The team shaped the river meanders according to Leopold and Wolman's equations; they sized the river according to the equations of hydraulic geometry. River restoration was growing up.

Other river restoration consulting firms were starting up at about the same time, first around the trout streams of the West and eventually throughout the United States. The trajectory of restoration in the 1990s essentially mirrored the trajectory of restoration in the early twentieth century, ranging from projects on the private lands of the Catskills to those on public lands for the CCC. As river restoration gained traction in the late twentieth century, projects took place less and less on private ranches or newly developed plots and increasingly on public lands like National Forests or state and county parks. Federal and state regulations also increased demand for restoration: California streams were restored in response to new salmon restoration requirements under the Endangered Species Act, and new surface mine reclamation laws required rebuilding and restoring streams in Wyoming. Federal agencies picked up on the trend as they sought to comply with environmental regulations: in South Florida, the Corps of Engineers began re-meandering the Kissimmee River, which they had finished de-meandering only thirty years earlier. There seemed to be no end of enthusiasm for stream restoration; by 2015, Inter-Fluve alone had restored over five hundred miles of streams. The number of stream restoration projects grew exponentially through the 1990s and into the twenty-first century, when over $1 billion per year was being spent on stream restoration nationwide.[1]

In the late 1990s, restoration evolved yet again, and firms like Inter-Fluve began to focus their gaze on dams. Ecologists and environmental scientists had long bemoaned the environmental destruction wreaked by dams. They stop flow. They frustrate salmon. They fragment landscapes. They kill rivers. John McPhee wrote emphatically in *Encounters with the Archdruid* that "there is something metaphysically sinister" about dams; that in the "absolute epicenter of hell sits a dam."[2]

Dams were—and are—rampant on America's rivers; in 2000, the Corps of Engineers estimated that the nation has over 80,000 dams more than ten feet tall. By that estimate, Americans have been building on average one dam per day since the signing of the Declaration of Independence. When we think of dams, we think of the Hoover Dam or Grand Coulee; but the vast majority of dams are small, private structures built decades or even centuries ago, often for reasons that are no longer relevant—from grinding grain for colonial villages to powering the mills of now-defunct textile manufacturers. Many of these obscure, ubiquitous dams are abandoned and useless for their original purposes, yet they are still causing all the harm that draws environmentalists' scorn.

By the close of the twentieth century, society was faced with painful end-of-life decisions for these geriatric pieces of infrastructure. In some cases, state agencies stepped in to pull the plug. Wisconsin and Pennsylvania, for instance, realized that old, abandoned dams were enormous safety hazards; locals called them drowning machines for kids. And so both states began to remove the most egregious cases of disused, harmful dams, and other states followed suit. Just as river restoration began quietly, as an almost secret passion of river eccentrics, so began the era of dam removal—slightly under the radar of national attention.[3]

———————

It's a strange thing to watch a dam being removed. In most cases it is anticlimactic, yet hypnotic. Like watching a building be assembled, but in reverse. A track hoe with a giant jackhammer noisily sidles up to the downstream side of the long concrete wall of the dam, which awkwardly protects its own destroyer. Then, for several hours, the jackhammer, arm extended, notches its way into the structure like a gigantic woodpecker. As work progresses, water begins seeping into and expanding the crack until some of the river is gushing through this narrow gap. Then the destroyer moves on along the dam to its next appointed notching. Soon the water no longer cascades over the breadth of the dam, but rather through the notches, eroding them further. Eventually the jackhammer is replaced with an enormous, scooping backhoe that shovels the bits and pieces of crumbled dam into a waiting dump truck, which unceremoniously carries the remnants to a landfill.

Two things are inevitable when removing a small dam. First, there will always be a crowd. Locals will come and watch, telling stories of things that happened at the dam. "That's where I proposed to Evelyn." "Remember when Billy's boy drowned?" Every dam is inexorably linked to its town and thus to its history. Second, as the dam is dismembered and the reservoir drains, there are always surprises. In Minnesota, for example, hockey nets and hundreds of hockey pucks gradually emerge. Often, as the concrete is removed, structural stratigraphic layers appear one by one. The hitherto unsuspected age and history of the dam are revealed as the underlying logs and jerry-rigged historic dam structure surface after a century of being blanketed with nondescript concrete.

Throughout the late 1990s, dam removal took place hundreds of times. Each time the work was similar, yet had a local flavor. In Wisconsin, adjustments had to be made for elephants from the Circus World Museum. For years these animals had bathed in the historic millpond of the Baraboo River; removing the dam left the elephants without their customary afternoon dip. In Oregon, restoring salmon runs by dam removal was a pet project of the Secretary of Interior Bruce Babbitt, who brandished a golden sledgehammer to take out the first notch. In North Carolina, to garner local support, dam removers brought in Marines from nearby Camp Lejeune with dynamite—river restoration as military training exercise.

From bulldozers to backhoes to explosives laid by helicopter, dam removal became part of stream restoration. Perhaps the capstone event of dam removal, when river use and abuse came full circle, took place on September 17, 2011, at the ceremony marking the removal of two dams on the Elwha River in the Olympic Peninsula of Washington state. In attendance were two senators, a governor, the commissioner of the Bureau of Reclamation, and the Secretary of the Interior—a group that likely would have been at the ribbon-cutting ceremony of a dam construction only a few decades earlier. The first notch was taken out of the dormant hydropower dam on the lower Elwha River. Backhoes floating on barges cut further notches while enormous cranes lifted house-sized concrete slabs from the top, working their way 257 feet down to the canyon floor, where swirling Chinook salmon bided their time.

Amid the fanfare—the politicians wielding golden sledgehammers and the giddy environmentalists—the economic accounting of dam removal and stream restoration more broadly was being tallied in the background. In Montana, digging out contaminated sediment and rebuilding Silver Bow Creek had not been cheap. Re-meandering a stream in Olympic National Forest was not easy. Removing a small dam was costly enough to show up as a line item on state budgets. Removing big dams, however, was downright expensive. Removing two dams on the Elwha River and restoring other stretches of the river ended up costing a shade under $325 million and was financed almost exclusively by the federal government.

If dam removal—or river restoration generally—was to occur at the scale needed to be effective, if it was going to affect a significant portion of the hundreds of thousands of miles of America's degraded streams and rivers, then the economic calculus would have to change.

Somehow, a different set of motives and mechanisms would have to come into play: the power of the market.

Much of the stream restoration work of the early and mid-twentieth century took place during a particular period for the national economy. The idea behind the funding of river restoration was that the economy should be supported by government spending on public goods, such as stream restoration. The economic growth after World War II, along with the enduring legacies of the Progressive Era and the Great Depression, had established the federal government as the economic engine of natural resources conservation. Stream restoration projects were often designed and built by private consulting firms like Inter-Fluve, and at times they were paid for by private ranchers; but more often, they were bankrolled by a federal agency or perhaps a state or local conservation program. If the problem of beat-up streams was going to be fixed, it was presumed that the government would have to do it. There were, however, a few divergent thinkers who questioned this assumption about the economy of environmental restoration.

John Dales is not your typical visionary. For starters, Dales was an economist—a practitioner of the dismal science. As a professor at the University of Toronto, sitting on the shore of Lake Ontario through the mid-twentieth century, Dales must have thought things looked particularly dismal. The rampant pollution of the Great Lakes and the surrounding rivers and streams had reached its nadir, and the burning Cuyahoga River was the poster child for all that had gone wrong. And yet there Dales was, quietly transforming our ideas of how to make it better.

Five hundred miles to his west, on the shore of Lake Michigan, Dales's more outspoken colleagues at the University of Chicago—Milton Friedman, Friedrich von Hayek, and Ronald Coase—were doing their best to reshape the world economy. They had unrestrained faith in the market, believing that minimal government combined with open markets was the most efficient approach to solving any policy conundrum. Poor health care? Privatize. High electricity prices? Deregulate. Underperforming schools? Vouchers. This rethinking of how the economy and government would fit together was unprecedented, and it notably proposed minimizing the role of government in the economy altogether.[4]

Dales didn't disagree with the Chicago boys, but he didn't totally agree either. He shared a basic faith in the market; but he was also interested in solving immediate problems, which called for pragmatic solutions to messy situations. In classic academic fashion, he sat down and worked out a thought experiment for how markets and prices might be used to reduce pollution in the Great Lakes, which today we call "cap and trade."

First off, Dales assumed that the public could not be counted on to address pollution. Instead, the government would need to recognize pollution as a public problem and mandate that the Great Lakes needed to be made cleaner by reducing it. By mandating that the Great Lakes have clean water, against the background of egregious pollution emptying into the lakes, the government would be creating a scarcity of clean water. Secondly, Dales realized that a market must have something to trade—a commodity. He argued that government should create this commodity—the right to emit pollutants. Each polluter—such as a steel manufacturer or a city's wastewater treatment plant—would be given a number of pollution permits and could either use the permits themselves or sell them to other polluters. If a polluter reduced its pollution sufficiently, it would have extra pollution permits to sell for profit. If its pollution increased, the polluter would then have to buy permits from someone else. Government agencies interested in improving water quality could reduce the number of permits given out over time, thus constraining total pollution and increasing the scarcity of the permits. In this way, the government could improve water quality by harnessing Adam Smith's idea of the invisible hand. Thus the pillars of free-market environmentalism—cap and trade, carbon trading, and biodiversity offsets—were all born in 1968 in Dales's slender 111-page book.[5]

Ronald Reagan implemented much of the ideology proposed by the Chicago boys. But applying market ideas to the environment—and especially to rivers—fell to his successor, George H. W. Bush. Bush's most significant adoption of free-market environmentalism was a series of amendments to the Clean Air Act, which applied John Dales's thought experiment to the real-world problem of the acid-rain-causing pollutants emitted by industrial smokestacks. Polluters were given permits for a certain quantity of pollution and told they could use their permits or trade them. The program worked shockingly well; air pollution was drastically reduced at a fraction of the expected cost.

Less visible was Bush's subtle application of markets to streams and wetlands through an obscure program known as "no net loss." Calls for no net loss started in the mid-1980s with the widespread realization that America's wetlands were rapidly disappearing. Since the turn of the nineteenth century, wetlands had been drained or filled—lost—at the rate of over sixty acres per hour. Wetlands conservation had become a surprising environmental plank in Bush's 1988 campaign. In 1989, as president, he followed up on that issue in a speech at a Ducks Unlimited meeting, where he asserted that the simple policy approach of no net loss would be an environmental legacy for generations to follow. In the nuance of regulatory language, streams and rivers were considered a special class of wetlands, and so the principle of no net loss applied to them as well.[6]

"No loss" of streams would have taken form as a cap—a ban on any development that happened to damage a stream or river. The inclusion of "net" was essential; it let in Dales's ideas. No net loss created a scarcity of conserved streams, along with a commodity that would enable possible stream markets to operate. If developers wanted to move or channelize 300 feet of a stream on a property, they simply had to ensure that 300 feet of stream elsewhere was restored to compensate; there had to be a net balance of natural streams. Destroying streams, or polluting them, or filling them, or running them underground was not illegal; it just required a permit by the Corps of Engineers, the regulatory agency charged with overseeing the no net loss program. The permit approach built in the flexibility for government to be responsive to the changing demands of society. The requirement for a permit meant that the government had a say in where wetlands and streams were lost, and the discretion to evaluate those permits according to shifting priorities. Making Dales's thought experiment real, the government created a cap-and-trade program for stream ecosystems.

In 2006 alone, the Corps evaluated approximately 96,500 permit applications under no net loss. That was the demand side. But this new market also required a supply side; that is, a supply of restored streams to be sold to developers in search of wetland and stream credits. This situation created an unusual business opportunity—a whole new market within an infant economy.

"We're in the business of swapping swamps." This is George Howard's summation of the whole stream-trading business and of his particular firm—Restoration Systems. Like many other enviropreneurs, George Howard and his business partner John Preyer got into stream trading without much idea if it would actually work.

Neither Howard nor Preyer are scientists; neither has spent much, if any, time on rivers or in labs like Doug Thompson's. But what they don't know about river science they make up for in spades with their knowledge of business and policy; they can plumb the absolute depths of any topic related to water policy and environmental markets. In 1996, when they founded Restoration Systems, environmental markets seemed like they had a big future, but there was no end of unknowns involved in entering such a nascent market.

Howard and Preyer met in college in the late 1980s. They both ended up in Washington, D.C., as congressional staffers at about the time that no net loss was simmering in federal policy circles. Over beers in D.C. bars, they guessed that if and when no net loss was given some teeth and made a requirement for all development projects, then dozens—maybe even hundreds—of land developers would need stream restoration projects for their developments to get permits from the Corps. The first people with stream restoration projects in motion and credits stored up would have a corner on the market.

And so Howard and Preyer were among the first people to put together a bank of environmental credits. As Preyer describes the new, emerging market, "At first, developers did the restoration themselves; they simply set aside a portion of the project property to move streams, did some token restoration, and that's it."

Howard finishes the thought: "But what developer wants to deal with environmental mumbo jumbo? They want their permit from the Corps to develop, and they don't want to give up any land if they can help it, so they'd rather just pay someone to make this whole environmental problem go away. That's where we come in."

The restored streams Howard and Preyer produce are formally called mitigation banks. And the commodity that Restoration Systems banks is called a stream credit. In North Carolina, where Restoration Systems was founded, one linear foot of restored stream equals one stream credit. Thus, when a developer impacts 300 feet of stream, that developer is debited 300 credits by the Corps of Engineers. That is, the developer owes the Corps 300 stream credits from somewhere, somehow. When the no net loss policy first went into practice, a developer simply would have restored 300 feet to generate the needed 300 credits, or found someone to restore 300 feet of stream on the developer's property. But before long, stream restoration firms like Restoration Systems began banking credits by restoring streams on land the company bought itself. Once the Corps approved a mitigation bank and certified it as a restored stream, the bank could sell its credits to developers, road builders, and whoever else needed credits for their projects but didn't want to deal with the hassle and expense of restoring streams foot by foot themselves.

In 2005 the going rate for a stream restoration credit was just above $200 per foot—a million dollars a mile. Stream restoration began to have the characteristics of a real market. Venture capitalists backed new mitigation banks. Federal and state agencies carefully regulated and monitored the market. Entrepreneurs leased or bought land, restored streams, and sold their credits to a wide range of buyers. And across the landscape, streams were tweaked and twisted, filled with woody debris, loaded with gravel, and scoured for fish to show their success. Mitigation bankers even had a trade group that lobbied state and federal legislators.[7]

But more and more bankers were getting into the business. An emerging army of engineers, hydrologists, and ecologists formed mitigation companies, competing to find available land on which to restore streams. Everything that John Dales had speculated about and that Milton Friedman had promised seemed to be happening; there was competition, and there was environmental restoration being done by private industry. With multiple mitigation banks in an area, developers soon had several options to choose from for credits. With competition came lower prices. With that, the aquatic dream of John Dales was realized: a functional market for streams, created by the government.

The bulldozers financed by mitigation bankers did largely the same type of work that had been done over the past decade in Oregon and Montana, reshaping channels into meandering rivers with pools and riffles as per the equations of Leopold and Wolman. With multiple companies doing similar work in similar areas, competition drove down prices. In North Carolina, one of the most competitive and active states in the restored stream market, it was difficult to get the price below $200 per foot of stream. Doing the work of restoring streams, regardless of the length of channel restored, seemed inevitably to require real estate attorneys, engineers, designs, and heavy machinery. Banking river credits had been a lucrative business, but it was quickly becoming a marginally profitable one.

In 2002, George and John from Restoration Systems saw an opportunity to innovate. Actually, they saw thousands of opportunities. Their home state of North Carolina had thousands of what the state called "functionally obsolete" dams. For George and John, all of these old, abandoned, and obsolete dams presented an unusual opportunity: they were environmentally damaging and a persistent public safety hazard, not to mention a financial liability for whoever happened to have inherited one of them. The dams of greatest interest to George and John were very old ones that had been abandoned when the textile industry left North Carolina decades earlier. Dams like Carbonton, which had been impounding the Deep River for well over a century. In fact, William Tecumseh Sherman had for some reason left the Carbonton Dam untouched while he blazed a path through the South in the closing days of the Civil War. The dam was eventually converted to power a textile mill that was itself eventually abandoned, after which the dam sat idle and obsolete for decades. George and John had seen the dismantling of dams paid for by government agencies and even a few removed by power companies that could no longer profitably produce power from them. They knew dam removal was possible; they wanted to see if it could be profitable.

The Deep River is aptly named. Relative to the rest of central North Carolina, it cuts a deep valley through the rolling agricultural landscape. On the cool morning of October 20, 2005, fog had settled in, filling the river valley. From the bridge, the river below was completely shrouded in mist; just a sinuous gray cirrus cloud snaking through the landscape. Eerie sounds rose from it—a persistent clanging and banging, as though trolls were at work below in the bowels of the earth. Occasionally the fog would clear and reveal first a glimpse of a yellow backhoe and then the wall of the dam with a gaping hole at its middle. Gradually, grudgingly, the Deep River was rerouting itself from the historic spillway of the dam toward the middle of the channel. As fog again engulfed the scene and the hidden sounds of restorative destruction echoed up the valley, George Howard stood on the riverbank with an enormous smile on his face. He had found a way to innovate river restoration.

The novelty—the innovation—was how simple it was to restore the river. The creative destruction planned by Restoration Systems restored over five miles of stream in one fell swoop, creating an enormous bank of stream credits for less than $70 per foot. These credits could easily be turned and sold on the market for $200 per foot—a tidy profit. George and John had just changed the calculus of stream restoration markets. Within five years after the Carbonton Dam was removed, two endangered mussel species had reestablished a home where the Deep River ran through the former impoundment. Fish were regularly migrating to and fro through the area, as they had been unable to do for over 150 years. The Deep River took its straight path through the impoundment and started to wiggle it; meanders began to form on their own, and sandbars, pools, and riffles appeared—just like Doug Thompson said should happen.

This work of removing a dam was a step forward in comparison to the hundreds of millions of dollars of federal funds needed for dam removal and restoration on the Elwha River. Like the other more typical stream restoration projects sprouting up on the landscape, this work was being funded by credit-seeking land developers. In this new restoration economy, the real estate market was setting the demand for restored streams. The Sunbelt states of the Southeast were undergoing a building boom in the early 2000s. Along with buildings came roads. Each time a road crossed a stream, the road builder had to buy stream credits. From 2000 to 2007, nearly 227,000 miles of roads were built in the United States, creating enormous demand for credits. Almost a thousand mitigation banks were started to meet this demand, resulting in developers purchasing nearly $3 billion worth of stream and wetland credits between 2000 and 2007. By the numbers, the program looked like a resounding success.[8]

But a problem had developed as stream restoration shifted from private Montana ranches to credit markets. It was a problem that had arisen before, almost a century earlier, when Hubbs was doing his restoration work for the CCC: there was no clear user for the restored streams. In a few cases, there were genuine buyers for restored streams. For instance, when Greg Koonce designed a river for a rancher or neighborhood in Idaho, it mattered whether trout actually showed up in the stream. Stream restoration was advertised for its ability to deliver the conditions for trout, and so Koonce's success was determined by whether his restoration produced more trout for the private landowner. But when permitting purposes were the motivation for stream restoration, the commodity was just linear feet of stream, restored to a point that satisfied the Corps' arbitrary standards: a stream restored that no one genuinely seemed to want. Even if re-meandering, or even dam removal, worked for restoring the actual ecology of streams and rivers, the work at Carbonton Dam had still been done to create the fictitious commodity of stream credits for sale in the equally fictitious credit market. The "market" forces were the pure creation of government—a purely political economy— little different perhaps from the government doing the work itself during the Great Depression.

The key question for restoration, and the true test of Adam Smith's philosophy, was whether anyone would take on restoring streams for purely self-interested motives without any government intervention or requirements.

In the high desert of eastern Oregon, cattle branding is a traditional community event, comparable to a barn raising back East. Neighbors—a relative term when the nearest ranch may be dozens of miles away—exchange help to get their annual branding, castrating, and vaccinating work done.

Scott Campbell runs his branding the old-school way—with neighbors and hired cowboys lassoing calves out of the herd. Cowboys on horseback plunge into a herd of cattle and emerge with a calf, roped at the head and feet. A dozen dusty hands leap on the wide-eyed calf, do their necessary deeds, and in less than three minutes the calf is back with mom. The event has an air of managed chaos; it's a cacophony of mooing cows and bleating calves, twirling lassos, and frantic cowhands dashing around pulling red-hot brands out of the coals of an open fire. It is hot, dusty, and hectic, yet also functionally elegant: an efficient system for branding that has evolved and been adapted to the realities of Western desert ranches over the course of generations.

Scott Campbell likes systems that have evolved over long periods of time, and the system that he believes is saving his ranch begins with saving the streams. There aren't many streams in this region. Campbell's ranch—Silvies Valley Ranch—is perched at an elevation of over 5,000 feet in eastern Oregon. A few of the hills have ponderosa pine, but the semiarid landscape is otherwise a barren stretch of dusty sagebrush occasionally broken up by rocky outcrops. Those streams that do exist are typically little more than occasionally moist gullies, resembling nothing more than trenches, incised deep into the silty valley sediment. There might be a narrow belt of grasses, or even willows here and there, but these are often settled deep within the gully, removed from the broader landscape by high streambanks. Above them, yet more sagebrush grows. This is the alluvial landscape of the high desert of the West, stretching from eastern Washington down to Arizona.

Cattle can't live on sagebrush; so ranchers have to irrigate to grow grass, if they can, or else buy hay. Both cost money. In a drought year like 2013, hardly any ranchers can grow enough forage to feed their herds. So they face buying hay—in a market where demand is driving the prices up rapidly—or selling much of their herd. With all the ranchers facing the same climate, a drought can lead to a massive sell-off of cattle across the region, and cattle prices inevitably plummet.

This arid landscape has a few unusual features. One of them is near a beaver dam, on Forest Service land adjacent to Silvies Valley Ranch. The view upstream from the dam is striking: surrounded by an expanse of drought-stricken, sagebrush-ridden landscape sit forty acres of lush green grass. "This is what this region would have looked like with beavers," is Campbell's perfectly reasonable interpretation. And to a rancher, a patch of green grass like this one is as good as gold.

Campbell grew up in this valley and is the son of the town doctor. He went to Oregon State University, became a vet, and then went big by seeing systems where others didn't. As a vet he started a pet hospital that grew into Banfield Pet Hospitals. He and his wife, Sandy—who also grew up in the high desert of eastern Oregon—built their first hospital into a franchise of 450 pet clinics. When PetSmart Stores began putting vet clinics in their stores, those were Banfield clinics.

During the astronomical growth of the Banfield business, the Campbells would return to a cabin in the Silvies Valley for vacations with their two sons. The cabin sits on a bluff, and below the bluff runs Hay Creek—a creek just like the myriad other gullied streams of the area. But in 2003, Campbell decided to work on the stream to give his sons somewhere to fish. Redband trout are the fish of choice here, and they are generally found only in the most pristine valleys of preserved lands. Campbell didn't know about Reds Wolman and Luna Leopold. He didn't hire engineers or geomorphologists. He just looked around, scratched his head, and got to work building rock weirs—dams about the same height as the streambanks—across the creek and floodplain. The idea was to slow down the water and create pools, and then the stream would flow across the weirs and create a small riffle, thus making some kind of habitat for redband trout. It worked, and Campbell's sons spent the next couple years pulling trout out of Hay Creek during summers at the cabin.

What Campbell didn't appreciate at the time was that his little rock dams on Hay Creek were replicating the work beavers had done before the West was settled, when fish and grass were, if not abundant, at least far more plentiful than they are today. In those days, beavers thrived, and their dams created streams that were little more than stairsteps of dams, ponds, and meadows. When a beaver builds a dam, the upstream pond of slowed water gradually fills with sediment. The downstream dam holds the water level just high enough to keep the ground slightly wet, grasses begin growing, and the pond becomes a grassy meadow with a narrow stream flowing through it. The natural landscape of the western high desert, with the prolific beavers doing their hydraulic engineering work, was a patchwork

quilt of grassy meadows and burbling brooks in the valleys and sagebrush on the uplands, rather than the plain blanket of sagebrush and
negligible damp gullies that now dominates.

Europeans created this existing landscape, foremost by removing
beavers. The first settlers on the western landscape were trappers,
who took the beavers with extraordinary thoroughness. In response
to the European craze for beaver-pelt hats, explorers and trappers
moved through eastern Oregon and southern Idaho in search of beaver. They removed 18,000 beavers from the area during 1823–1829
and took thousands more in the 1830s. But by the 1850s, trappers
were scraping to find a few hundred animals. Thanks to the trans-
atlantic demand for beavers, they were virtually eradicated. Without
benefit of constant maintenance, beaver dams throughout the American West were lost, along with all of the lush grass and trees in the
valleys. As the dams decayed and were washed away, the sediment-
laden meadows were eroded, leaving behind a lunar-like landscape of
barren gullies.[9]

When Campbell bought the entire Silvies Valley Ranch in 2007,
he gradually became more aware of this history of beavers. He
found remnant beaver dams and meadows in the region, and he
noticed references to beavers in old descriptions of his property. As
he got the ranch back into working order by rebuilding barns, roads,
and hundreds of miles of fences, he took on the gullied streams on
the property. Because Campbell already had backhoes and bulldozers working throughout the ranch, he figured he might as well fix up
a few of the creeks on the ranch for trout. "Yeah, initially I was in
it for the fish," he chuckles as he walks along the recently restored
Flat Creek.

On this creek, and on others at the ranch, Campbell is taking the
approach he used years earlier: building a series of rock weirs across
the gullies, spacing them along the valley just often enough to form
artificial beaver ponds and meadows. And it has worked. Fish are
now present in the restored Flat Creek, but the grass is the most
impressive. Campbell is doing the beavers' work in Silvies Valley. The

rock weirs he built across the streams create habitats for trout, but they also back up water in the stream. When streams are trenched and incised, spring rain and snowmelt quickly flow downstream and the floodplain soils are perched too high for vegetation to capture any of the shallow groundwater. The rock weirs slow down the flow enough to saturate the soils, and the groundwater is backed up so that it flows very slowly, just beneath the surface of the silty flood-plain soil. As the groundwater rises, the sagebrush dies off, unable to survive in such moist soils, and native grasses spring up. No planting needed; just add water. When one of the ranch's cowboys describes going by Flat Creek a year after the restoration, he says, "I was walk-ing around and it was all green. And it wasn't just grass; it was hay."

The most gripping experience at Flat Creek is listening to the sounds. Water gurgles over the weir, trout and otters splash as they surface, and what seems like countless birds are chirping and whis-tling in and out of the area as breezes rustle the tall grass. Camp-bell has done the impossible. He did what the Willowemoc Club, the CCC, the University of Michigan, the Corps of Engineers, and countless engineers weren't been able to do. More than simply restor-ing a stream, he restored an ecosystem.

Scott Campbell's streams are unlike almost any other restoration project. Stream restoration in all corners of America has become as complicated as open heart surgery; heavy construction equip-ment proliferates, spewing diesel smoke as restorationists flay open riparian corridors and cut down riverfront trees to make way for backhoes and bulldozers. The whole process is guided by equations and computer models that will lead to reconditioned, sinuous, pre-sumably restored streams. After all of this work, most stream resto-ration projects are successful only by declaration. Restored streams are inevitably marked with a respectable sign emblazoned with the logos of the agencies and firms involved in the work, the obligatory billboard announcing ecological victory. But aside from the signs, evidence for ecological success of most stream restoration projects can be hard to find.

Scott Campbell has gradually learned all this. He attends restoration conferences and reads all of the literature. He has come to
think it is all crazy. Van Cleef and Hubbs, along with countless other
engineers, had built structures in streams based on the thinking that
they were the key missing ingredient. Campbell has taken a big step
back and guessed that what was wrong in the stream was a symptom
of a broader problem in the landscape and that beavers might be
the missing ingredient: restore the beavers, he reasons, or at least
the effect of the beavers, and the streams will follow. Campbell's
approach is radically out of the ordinary, deliberately different from
what others are saying should be done. And it works.

As importantly, the strategy works for Campbell as a rancher,
who has to think in terms of animal unit months (AUMs)—that is,
the capacity amount of feed—to graze a cow for a month. Typically,
in this area of Oregon, Campbell needs five acres per AUM, which
he can either use himself or lease to another rancher for $25 for the
year. Therefore, typical grazing land is worth $5 per acre per year.
Along Flat Creek, before he started working on it, there were 600
acres of riparian grazing land. Assuming five acres per AUM with
typical grazing in this semiarid land, and $25 per AUM, Campbell
could count on the Flat Creek meadow grazing land being worth
roughly $3,000 per year.

In 2013, three years after Flat Creek was restored, the math has
changed. He now has 3 AUM per acre instead of 5 acres per AUM.
At $25 per AUM, Flat Creek meadow now generates $45,000 per
year as grazing land. Furthermore, as the cowboy had noted, this land
is growing hay. Campbell now gets a cutting of hay off the meadow
and can still put cattle out to graze afterward. He is now baling from
1 to 1.5 tons of hay per acre from the 600-acre meadow, and hay is
netting over $120 per ton, leaving Campbell with over $70,000 worth
of hay that he can either sell or use for his own cattle. In 2013, when
the entire region was in for a long drought and most ranchers were
faced with either buying hay to support their herd or selling it off
at reduced prices, stream restoration helped Campbell ride it out.

As he captures it, "I've been able to increase my production in these restored areas by a factor of ten. And I'm getting the trout too."

All of this success is a return on an investment. And for Campbell, keeping the investment side of the equation down is just as important as getting results. He spent about $100,000 per mile on restoring 1.5 miles of Flat Creek and over 3 miles of nearby Camp Creek. But Campbell and his hired hands are always looking for ways to get the work done for less money. Steve, a soft-spoken ranch hand at Silvies Valley who did a lot of the actual construction work on the streams, gives a sly grin when he nods at the work being done on another stream on Campbell's ranch, "We've got it down to about twenty thousand a mile now."

Though Campbell knows that other restoration efforts are investing in re-meandering and planting willows, for him, "The return would have to be massive for that level of investment." Instead, Campbell has been doing as little as possible to generate positive returns. He guesses that most of his stream restorations will pay for themselves in two to three years. But he is always quick to point out that he is getting the hay, the grazing, and the increased land value, all in addition to what he was initially after: the trout. He is getting what all the other restoration projects are after; but by doing it differently, he is getting a whole lot more for a whole lot less.

Most people involved with stream restoration don't fool themselves into thinking it is an economically sustainable enterprise. At $1 million a mile with unknown ecological benefits at a time of shrinking government coffers or housing bubble collapses, it's doubtful that such environmental political experimentation can—or should—be sustained decades into the future. But in the dusty ranchlands of eastern Oregon, Scott Campbell may have developed a true restoration economy.

Other ranchers are watching Campbell's restoration work not just because his approach is new, but because it genuinely seems to work. A few ranchers were really irritated at first, but in time they have come around to see the benefits. Some of the ranchers at the branding are starting restoration projects on their own lands. Over a post-branding lunch of elk steaks and potato salad, a neighboring rancher begins to describe why he and other ranchers are interested in stream restoration. Why are they voluntarily doing what the government has to require people to do back East? The first reasons he gives are what might be expected: a sense of pride, and the truism that good ranchers are always thinking about how their ranch will be not just next year, but in fifty years. These are the pat answers explaining why ranchers and farmers should be expected to manage their lands sustainably. The answers have been used many times before, often unconvincingly to outsiders. But then this rancher runs his cold blue eyes across the vast, open horizon—in the distant way that only cowboys seem able to do—and says something profound about restoration in the ranches of the West: "The big reason is *political* sustainability. Most ranchers just want to be left alone to do their work, be with their family, and ranch. If the public is aware that ranchers are managing their lands better than agencies would, then there's a better chance ranchers will get left alone. That—being left alone because people trust you're doing what's right—that's a big part of sustainability for western ranching."

And it may just be that Scott Campbell figured out a way to make some ranches in the semiarid West more sustainable. In the beginning, Campbell was no different from the vacationing trout fishermen of the Catskills—just a wealthy businessman trying to make trout fishing better at his weekend cabin. But his discovery that stream restoration was central to restored grasslands for his cattle took the problem from raising more trout to having a healthier ecosystem. His streams of trout had led to streams of grass and restored an economy in the process.

ACKNOWLEDGMENTS

People who work on and around rivers love their jobs. They also love telling people about their jobs, which is what made this book possible. I mention many of these people directly, but others played even more important educational and logistical roles behind the scenes.

Cap'n Bill Beacom cuts a powerful wake across the Midwest; his arrangement of my trip on the *Christopher Parsonage* down the Mississippi River was a favor I can never repay. Beyond that, Bill answered countless questions with his bottomless knowledge of anything relating to rivers and barges. The crew of the *Parsonage* included some of the most savvy river scientists I have ever met—none more so than Howdy Duty and Donnie Randleman, who patiently tolerated me in their wheelhouse.

Joe Gibbs and Susan Dell'Osso taught me about levee districts, Daryl Armentrout and Zyg Plater about the TVA, Patrick Malone about textile mills, Ben Copp about cattle and grazing, Adam Riggsbee and Todd BenDor about mitigation banking, and Marty Reuss about the history of the Corps of Engineers. Brian Chaffin guided me through the Klamath by raft and by car, and I would have far less appreciation for tribal water without his scholarly work and his enthusiastic guiding. Chuck Podolak and Disque Deane did their best to educate me about western water and the Colorado River, while Jeff Muehlbauer and Ted Kennedy took me on a raft down the river to make sure I learned it as well as a southerner could. The finest ranch in the United States is, hands down, Silvies Valley Ranch, where Sandy Campbell is one of the kindest and most generous people I've met.

I had the great fortune to spend a year at the Institute for Water Resources at the Corps of Engineers as a Frederick Clarke Scholar. While there I acquired enormous institutional knowledge of navigation and flood control, along with some time to start digging into the history of the Corps and the opportunity to pester Gerry Galloway with many questions. In addition, Luna Leopold patiently corresponded with me about the early days of geomorphology while Reds Wolman regaled me with similar stories during my visits to his office; I regret that I didn't finish the book before both of them passed away.

A Guggenheim Fellowship allowed me the intellectual space to start writing, while the librarians at the Lake Pleasant Public Library on the banks of Lake Pleasant, New York, gave me the literal space to write much of the first draft.

My incipient career in rivers was set in motion by Doug Shields amid long days in the Mississippi delta, learning about Humphreys, Twain, Muddy Waters, and Faulkner while also learning about shear stress and hydraulic habitat. From there, I was guided by the first-class scholars Jon Harbor and Emily Stanley. I had the great fortune of spending almost a decade in the geography department at the University of North Carolina, where I benefited by endless sidewalk conversations about water and history with Larry Band, John Florin, Steve Birdsall, and Chad Bryant.

My riverine friends and colleagues elsewhere—particularly Rebecca Lave, Morgan Robertson, Frank Magilligan, Lauren Patterson, Jack Schmidt, Robb Jacobson, and Will Graf—have been powerful resources for all things related to rivers and society. Beaver meadows and dam removal gave me an excuse to know Gordon Grant, and I've benefited tremendously from that friendship.

My PhD students for the past fifteen years have been, without exception, extraordinary. On top of their own scholarly abilities and accomplishments, they have shown exceptional patience with my endless historical distraction and tangents.

Matt Weiland and Don Lamm—my publisher and agent—have made me as much of an author as possible, overcoming my shortcomings with their skills. Remy Cawley's surgical editing was necessary, to say the least. Some of my colleagues provided invaluable reviews of earlier versions of chapters—Jim Salzman, Rebecca Lave, Jeff Mount, Will Graf, Ben Copp, and Brian Chaffin.

My parents took me canoeing on the Buffalo River of Arkansas in a great big yellow canoe—an experience that set my life in the right direction. They then pushed me to write and rewrite. My brother, Wyatt—also a river scientist—has taught me the Missouri River, and river fisheries, as only a big brother could.

My kids have spent many hours with me pulling oars on Lake Pleasant, rolling a kayak on the Sacandaga, and tipping rocks on the Eno. Whether they enjoyed it or just tolerated it, I appreciate their companionship just the same.

Finally, twenty years ago, Carrie Blomquist canoed the Upper Missouri River with me. Since then she has humored my inordinate enchantment with rivers, hopefully knowing that it is surpassed only by my fascination with her.

ILLUSTRATION CREDITS

NOTES

Chapter 1: Navigating the Republic

1. Jefferson to Washington, 15 March 1784, in Thomas Jefferson Papers, Library of Congress *American Memory* website, https://memory.loc.gov/. Note that the spelling of *Potomac* has changed through the centuries, and I have used the current preferred spelling throughout the quotes.

2. A great, readable account of Washington's trip through the region is summarized in J. Achenbach, *The Grand Idea: George Washington's Potomac and the Race to the West* (New York: Simon & Schuster, 2004), 47–120; the quote is at p. 114.

3. H. Adams, *Life of Albert Gallatin* (Philadelphia, PA: J. B. Lippincott, 1879), 56–58.

4. Washington to Jefferson, 29 March 1784, George Washington Papers, Library of Congress *American Memory* website.

5. The three men would eventually go on to be formative in the government they helped construct: Hamilton was the first Secretary of Treasury, Madison served in the House of Representatives in the first Congress, and Jay was the first Chief Justice of the Supreme Court.

6. John Jay, *Federalist* No. 2, in *The Federalist Papers*, ed. C. Rossiter (New York: New American Library, 2003), 32.

7. Alexander Hamilton, *Federalist* No. 11, in *The Federalist Papers*, 79.

8. With the inclusion of the Bill of Rights in 1789, unless a function was explicitly given to the national government, that function was the responsibility of the individual states.

9   C. C. Weaver, "Internal improvements in North Carolina previous to 1860," *Johns Hopkins University Studies in Historical and Political Science* 21, no. 3–4 (1903): 121–27, 161–79.

10   A. Parkman, *History of the Waterways of the Atlantic Coast of the United States* (Alexandria, VA: U.S. Army Corps of Engineers National Waterways Studies, 1983), 20–25.

11   G. R. Taylor, *The Transportation Revolution, 1815–1860* (New York: Rinehart, 1951), 32–55.

12   Parkman, *History of the Waterways of the Atlantic Coast*, 33–34.

13   J. F. Stover, *American Railroads* (Chicago: University of Chicago Press, 1997), 6–7.

14   P. L. Bernstein, *Wedding of the Waters: The Erie Canal and the Making of a Great Nation* (New York: W. W. Norton, 2005), 304.

15   Packet boats on the Erie Canal typically had three horses, carried no freight, and traveled quickly. Line boats cost less (about two-thirds less), had only two horses, and were much heavier, but they traveled more slowly. Letter from Karl Brunnhuber at Erie Canal Museum in Syracuse, New York; additional descriptions of traveling on the Erie Canal in this era can be found in F. Deoch, *New York to Niagara, 1836: The Journal of Thomas S. Woodcock* (New York: New York Public Library, 1938).

16   Bernstein, *Wedding of the Waters*, 353–55.

17   Details of revenue, expenditures, and economic status of the Potomac Canal from R. J. Kapsch, *The Potomac Canal* (Morgantown: West Virginia University Press, 2007), 242–52.

18   "Valuation, Taxation, and Public Indebtedness, VII," *Tenth Census of the United States, 1880* (Washington, DC: U.S. Government Printing Office, 1884), 523–26.

19   James Madison, *Federalist* No. 41, in *The Federalist Papers*, 254–55.

20   Alexander Hamilton, *Federalist* No. 23, in *The Federalist Papers*, 149.

21   Hamilton, *Federalist* No. 30, in *The Federalist Papers*, 184.

22   Hamilton, *Federalist* No. 24, in *The Federalist Papers*, 157.

23   Shallat provides a thorough history of the formative years of the Corps of Engineers and its ontogeny; see particularly T. Shallat, *Structures in the Stream: Water, Science, and the Rise of the U.S. Army Corps of Engineers* (Austin: University of Texas Press, 1994), 30–42, 79–100.

24   K. Baumgardt, "Robert E. Lee: A Personal Look at Baltimore's 'First' Engineer," in *Baltimore Civil Engineering History*, ed. B. G. Dennis and M. C. Fenton (New York: American Society of Civil Engineers, 2005), 53–64.

25   Alexander Hamilton, *Federalist* No. 78, in *The Federalist Papers*, 465.

26   Alexander Hamilton, *Federalist* No. 80, in *The Federalist Papers*, 479.

27 W. F. Swindler, *The Constitution and Chief Justice Marshall* (New York: Dodd, Mead and Company, 1978), 77–86, 358–74.

Chapter 2: Life on the Mississippi

1 During some years in the 1830s and 1840s, the values of exports from New Orleans actually *exceeded* those of New York. G. R. Taylor, *The Transportation Revolution, 1815–1860* (New York: Rinehart, 1951), 7–9.

2 J. Mak and G. Walton, "Steamboats and the great productivity surge in river transportation," *Journal of Economic History* 32, no. 3 (1972): 619–40.

3 Mak and Walton, "Steamboats and the great productivity surge," 619–40.

4 These values and estimates were for the lower part of the Missouri River, a particularly nasty river to navigate. P. O'Neil, *The Rivermen* (New York: Time-Life Books, 1975): 83, 130, 144.

5 See generally Alexander Hamilton, James Madison, and John Jay, *Federalist* No. 35 and 36, in *The Federalist Papers*, 207–20.

6 Alexander Hamilton, quote from *Federalist* No. 31, in *The Federalist Papers*, 190.

7 F. G. Hill, *Roads, Rail, and Waterways: The Army Engineers and Early Transportation* (Norman: University of Oklahoma Press, 1957), 18–36, 163, 184–85, 195.

8 M. Twain, *Life on the Mississippi* (New York: The Modern Library, 1994), 212.

9 S. B. Carter et al., *Historical Statistics of the United States, Earliest Times to the Present: Millennial Edition* (New York: Cambridge University Press, 2006), chap. Df, table Df, 690–91.

10 National Research Council, *The Missouri River Ecosystem* (Washington, DC: National Academies Press, 2002), 10–11.

11 M. Grunwald, "An Agency of Unchecked Clout; Water Projects Roll Past Economic, Environmental Concerns," *Washington Post*, September 10, 2000. J. A. Hird, "The political economy of pork: Project selection at the U.S. Army Corps of Engineers," *American Political Science Review* 85 (1991): 441–50.

12 For the years 1925–1957: C. E. Landon, "Freight traffic on the Ohio River," *Financial Analysts Journal* (1961): 51–56. For years since 1975: U.S. Army Corps of Engineers, *Waterborne Commerce of the United States, Annual Data* (Alexandria, VA: Institute for Water Resources, 2015), table 3-2.

Chapter 3: The Rise of the Levees

1 W. Wilson, "The study of administration," *Political Science Quarterly* 56 (1941): 505.

2 For general evolution of federalism, see D. J. Elazar, "Opening the third century of American federalism: Issues and prospects," *Annals of the American Acad-*

emy of Political and Social Science 209 (1990): 11–21. Also see J. Kincaid, "From cooperative to coercive federalism," *Annals of the American Academy of Political and Social Science* 509 (1990): 139–52.

3   The organization and finances of levee districts have received limited historical attention. Harrison's work is the primary effort at analyzing available records in the Lower Mississippi River region. R. W. Harrison, *Alluvial Empire, Volume One: A Study of the State and Local Efforts Toward Land Development in the Alluvial Valley of the Lower Mississippi River* (Little Rock, AR: Pioneer Press, 1961), 89–92. A general overview with far less financial information is also available: R. W. Harrison, "Flood control in the Yazoo-Mississippi Delta," *Southern Economic Journal* 17 (1950): 148–58.

4   Many districts in the Upper Midwest and the Illinois River valley are referred to as "drainage districts" because they had the combined tasks of first converting flooded wetlands into farmable property and then keeping river water out. J. Thompson, *Wetlands Drainage, River Modification, and the Sectoral Conflict in the Lower Illinois Valley, 1890–1930* (Carbondale: Southern Illinois University Press, 2002), 10–11. For a description of levee construction at the Sacramento River, see R. Kelley, *Battling the Inland Sea: American Political Culture, Public Policy, and the Sacramento Valley 1850–1986* (Berkeley: University of California Press, 1989), 10–15.

5   Thompson, *Wetlands Drainage, River Modification, and Sectoral Conflict*, 16. I have corrected the misspellings that occurred in the original.

6   G. D. Lewis, *Charles Ellet, Jr.: The Engineer as Individualist* (Urbana: University of Illinois Press, 1968), 12–23.

7   C. D. Calsoyas, "The mathematical theory of monopoly in 1839: Charles Ellet, Jr.," *Journal of Political Economy* 57 (1950), 170.

8   J. M. Barry, *Rising Tide: The Great Mississippi Flood of 1927 and How It Changed America* (New York: Simon & Schuster, 1997), 33–37.

9   Lewis, *Charles Ellet*, 134–43.

10   The full title was even more intimidating: A. A. Humphreys and H. L. Abbot, *Report Upon the Physics and Hydraulics of the Mississippi River; upon the Protection of the Alluvial Region against Overflow; and upon Deepening the Mouths: Based upon Surveys and Investigations Made under the Acts of Congress Directing the Topographical and Hydrographical Surveys of the Delta of the Mississippi, with such Investigations as Might Lead to Determine the Most Practicable Plan for Securing it from Inundation, and the Best Mode of Deepening the Channels at the Mouths of the River* (Washington, DC: United States Army Professional Papers of the Topographical Engineers), 1876. Abbot quote is from M. Reuss, "Andrew A. Humphreys

and the development of hydraulic engineering: Politics and technology in the Army Corps of Engineers, 1850–1950," *Technology and Culture* 26 (1985): 9.

11   Humphreys and Abbot, *Report on the Physics and Hydraulics of the Mississippi River*, 192.

12   C. Ellet, *The Mississippi and Ohio Rivers: Containing Plans for the Protection of the Delta from Inundation; and Investigating the Practicability and Cost of Improving the Navigation of the Ohio and Other Rivers by Means of Reservoirs* (Philadelphia, PA: Lippincott, Grambo and Company), 1853.

13   M. Reuss, *Wetlands, Farmlands, and Shifting Federal Policy: A Brief Review* (Washington, DC: U.S. Army Corps of Engineers, 1994), 3–5.

14   Ellet, *The Mississippi and Ohio Rivers*, 28.

15   For the Mississippi and Missouri Rivers: U.S. Army Corps of Engineers, *Floodplain Management Assessment of the Upper Mississippi River and the Lower Missouri Rivers and Tributaries* (Minneapolis, MN: Minneapolis–St. Paul District U.S. Army Corps of Engineers, 1995), 2–6. For the Illinois River: Thompson, *Wetlands Drainage, River Modification, and Sectoral Conflict*, 59–60.

Chapter 4: Flood Control

1   For scientific expertise of levee district engineers: W. Starling, "Flood heights in the Mississippi River, with especial reference to the reach between Helena and Vicksburg," *Transactions of the American Society of Civil Engineers* XX (1889): 195–228. Probabilities for flooding went from roughly once every 3 years to once every 6 years, based on an inventory of Mississippi levee district crevasse data. R. W. Harrison, *Alluvial Empire, Volume One: A Study of the State and Local Efforts Toward Land Development in the Alluvial Valley of the Lower Mississippi River* (Little Rock, AR: Pioneer Press, 1961), 116–18, table III-5. Years include 1867, 1882, 1883, 1884, 1890, 1897, 1903, 1912, 1913, and 1927.

2   For impacts in the northeast and Midwest, see J. L. Arnold, *The Evolution of the 1936 Flood Control Act* (Alexandria, VA: Office of History, U.S. Army Corps of Engineers, 1988), 17–18. For Sacramento River, see R. Kelley, *Battling the Inland Sea: American Political Culture, Public Policy, and the Sacramento Valley 1850–1986* (Berkeley: University of California Press, 1989), 268–69, 277. In the late nineteenth century, the federal government tasked two commissions to begin addressing complaints about the lack of a federal role in flood control: the Mississippi River Commission and the California Debris Commission. Both groups were made up of appointed officers from the Corps of Engineers and charged with developing plans to reduce the flooding as a mechanism of

sustaining navigation. Neither commission was to do the actual work; rather, they were to provide planning and technical information that the states were responsible for implementing. Both groups were attempts to strike a balance between national government engagement and state government responsibilities; but their actions were always under the guise of navigation, in which the federal government had a clear role. M. D. Mitchell, "Land and water policies in the Sacramento–San Joaquin Delta," *Geographical Review* 84 (1994): 412–18.

3   The most compelling account of the 1927 flood is J. M. Barry, *Rising Tide: The Great Mississippi Flood of 1927 and How It Changed America* (New York: Simon & Schuster, 1997). For additional statistics on areas flooded and impacts, see Harrison, *Alluvial Empire*, 148–49.

4   Arnold, *Evolution of the 1936 Flood Control Act*, 18; C. A. Camillo, *Divine Providence: The 2011 Flood in the Mississippi River and Tributaries Project* (Vicksburg: Mississippi River Commission, 2012), 16–17.

5   M. Grunwald, *The Swamp: The Everglades, Florida, and the Politics of Paradise* (New York: Simon & Schuster, 2006), 191–96.

6   R. M. Smith, "The politics of Pittsburgh flood control, 1908–1936," *Pennsylvania History* 42 (1975): 5; Arnold, *Evolution of the 1936 Flood Control Act*, 58–65.

7   Federal restraint in flood control had begun to unravel with the first Flood Control Act in 1917, which authorized $45 million for the Mississippi River and another $5.6 million for the Sacramento River. While limited in geographic scope and continuing to focus on levees alone, the 1917 act was revolutionary in openly acknowledging flood control as a federal government interest. Its immediate effect was limited in part because levee districts had already accomplished so much: no work was done on the levees of the Yazoo–Mississippi Levee District following the 1917 act because their levees were already up to the specifications set by the U.S. government, and almost 70 percent of the Mississippi Levee District was already up to federal standards. Harrison, *Alluvial Empire*, 195, 226. The role of levees was also complicated due to the ongoing linking of levees with navigation; thus, a levees-only policy could be linked with keeping the Corps' role restricted to navigation for interstate commerce. C. A. Camillo and M. T. Pearcy, *Upon Their Shoulders: A History of the Mississippi River Commission from Its Inception Through the Advent of the Modern Mississippi River and Tributaries Project* (Vicksburg: Mississippi River Commission, 2004), 85–101.

8   To appreciate the growth in the size of levees, see J. D. Rogers, "Development of the New Orleans flood protection system prior to Hurricane Katrina," *Journal of Geotechnical and Geoenvironmental Engineering* 134 (2008), 615.

9   Camillo provides a detailed account of the 2011 flood from the perspective of the

Corps of Engineers leadership, including a particular chapter focused on the river in the region of Greenville, Mississippi. Camillo, *Divine Providence*, 164–216.

10 A number of studies have focused on the relationship between Congress and the Corps; Ferejohn's is particularly critical and analytical. J. A. Ferejohn, *Pork Barrel Politics* (Stanford, CA: Stanford University Press, 1974), 19–21, 25–46, 233–35. For Missouri River infrastructure, see L. Cartwright, "An examination of flood damage data trends in the United States," *Journal of Contemporary Water Research and Education* 130 (2005): 20–25.

11 National Research Council, *New Directions in Water Resources Planning for the U.S. Army Corps of Engineers* (Washington, DC: National Academy Press, 1999), 19–27.

12 A thorough review of the climatology and hydrology of the 1993 flood is given in S. A. Changon, *The Great Flood of 1993: Causes, Impacts, and Responses* (Boulder, CO: Westview Press, 1996), chaps. 2 through 4. For efforts to fight the flood, see G. A. Tobin, "The levee love affair: A stormy relationship," *Journal of American Water Resources Association* 31 (1995): 363. For the spatial and economic impact of the floods and comparison to 1927 flood, see M. G. Anderson and R. H. Platt, "St. Charles County, Missouri: Federal Dollars and the 1993 Midwest Flood," in *Disasters and Democracy: The Politics of Extreme Natural Events*, ed. R. H. Platt (Washington, DC: Island Press, 1999): 215–18.

13 Those in urban and regional planning circles refer to this situation as the "safe development paradox"; see R. Burby, "Hurricane Katrina and the paradoxes of government disaster policy," *Annals of the American Academy of Political and Social Science* 604 (2006): 171–91.

14 For the evolving role of FEMA and its expanding mission, see Platt (ed.), *Disasters and Democracy*, 16–18. Disaster relief has typically been associated with civil defense. During the Cold War it was assumed that disasters would be caused by military engagements; over time it became clear that natural disasters were chronic, and so FEMA was located in the office of the White House. Thus the return of FEMA to the Department of Homeland Security under the Bush Jr. administration should not be viewed as anomalous: R. Sylves and W. R. Cumming, "FEMA's path to homeland security: 1979–2003," *Journal of Homeland Security and Emergency Management* 1, no. 2 (2004): 11.

15 For the Stafford Act, see Platt (ed.), *Disasters and Democracy*, 23–26. For the number of disaster declarations, see M. A. Mayer, "Congress Should Limit the Presidential Abuse of FEMA," *Heritage Foundation WebMemo* No. 3466 (2012).

16 U.S. Senate Bipartisan Task Force on Funding Disaster Relief, *Federal Disaster Assistance* (Washington, DC: U.S. Government Printing Office, 1995), 15, 70.

17   J. Chivers and N. E. Flores, "Market failure in information: The National Flood Insurance Program," *Land Economics* 78, no. 4 (2005): 519–20.

18   Platt (ed.), *Disasters and Democracy*, 31.

19   The Galloway Report is formally known as the Interagency Floodplain Management Review Committee, *Sharing the Challenge: Floodplain Management into the 21st Century* (Washington, DC: U.S. Government Printing Office, 1994). Quote is from p. 180.

20   For discussion of moral hazard in the housing crisis, see P. Krugman, *The Return of Depression Economics and the Crisis of 2008* (New York: W. W. Norton, 2009), 63. For the importance of the Galloway Report in raising the idea of moral hazard in context of federal approach to natural hazards, see Burby, "Hurricane Katrina," 180.

21   Editorial, "Awash in Tax Dollars," *Raleigh News & Observer*, November 11, 1997; Platt (ed.), *Disasters and Democracy*, 40–41.

22   A. Kamen, "Nomination Stuck in the Mud," *Washington Post*, April 28, 1995.

23   For dimensions of drainage infrastructure, see Rogers, "Development of the New Orleans Flood Protection System," 612. The Saucer quote is from Douglas Brinkley, *The Great Deluge* (New York: William Morrow, 2006), 13.

24   The sequence of decision making for the flood control system at New Orleans dates back to the eighteenth century and is tremendously complex. The most thorough analysis of archival material, particularly the technical and engineering documents regarding the decisions made, was an investigation contracted by the Corps of Engineers. D. Woolley and L. Shabman, *Decision-Making Chronology for the Lake Pontchartrain and Vicinity Hurricane Protection Project: Final Report for the Headquarters* (Alexandria, VA: U.S. Army Corps of Engineers, 2007), chap. 4: 4-1–4-27. See also J. D. Rogers et al., "Interaction between the U.S. Army Corps of Engineers and the Orleans Levee Board preceding the drainage canal wall failures and catastrophic flooding of New Orleans in 2005," *Water Policy* 17 (2015): 707–23.

25   A. Carrns, "Long Before Flood, New Orleans System Was Prime for Leaks," *Wall Street Journal*, November 25, 2005.

26   Platt (ed.), *Disasters and Democracy*, 175–76.

27   Woolley and Shabman, *Decision-Making Chronology*, 2-49–2-50; M. Derthick, "Where federalism didn't fail," *Public Administration Review* 67 (2007), 39; University of California at Berkeley and American Society of Civil Engineers, *Preliminary Report on the Performance of the New Orleans Levee Systems in Hurricane Katrina on August 29, 2005* (Washington, DC: National Science Foundation Report No. UCB/CITRIS-05/01, 2005), 6-2.

28 The most thorough and exhaustive timeline of events during Katrina remains Brinkley, *The Great Deluge*. For a description of Coast Guard efforts, see pages 209–14 in that book.

29 D. Alexander, "From civil defence to civil protection—and back again," *Disaster Prevention and Management* 11 no. 3 (2002): 209–13.

30 Chertoff quote from Brinkley, *Great Deluge*, 634. For the role of the National Response Plan in slowing down response to disaster, see F. F. Townsend, "A Week of Crisis—August 29–September 5," *The Federal Response to Hurricane Katrina: Lessons Learned* (Washington, DC: Office of the White House, 2006), 33–50.

31 Limiting the use of federal troops in U.S. territory has been staunchly observed since the Posse Comitatus Act was passed in 1878.

32 A. M. Giuliano, "Emergency federalism: Calling on the states in perilous times," *University of Michigan Journal of Law Reform* 40 (2007): 351.

33 Brinkley, *The Great Deluge*, 413–15, 565. For Jeb Bush's response during Hurricane Wilma, see Derthick, "Where federalism didn't fail," 43–45.

34 This idea of the central role of experiment in America's government is well stated by T. Ferris, *The Science of Liberty: Democracy, Reason, and the Laws of Nature* (New York: HarperCollins, 2010), 101–5.

35 For an early description of states as laboratories of policy, see *New State Ice Company v. Liebmann*, 285 U.S. 262, 311.

Chapter 5: Water Wars

1 One of the most fascinating perspectives of the Klamath water wars is that told through the actions of the office of U.S. Vice President Dick Cheney. B. Gellman, *Angler: The Cheney Vice Presidency* (New York: Penguin Press, 2008), 195–213.

2 P. Bump, "That Time Ronald Reagan Joined a 'Rebellion'—but Still Couldn't Change Federal Land Laws," *Washington Post*, January 4, 2016.

3 W. Cronon, *Changes in the Land: Indians, Colonists, and the Ecology of New England* (New York: Hill & Wang, 1984), 69–71.

4 J. W. Dellapenna, "The evolution of riparianism in the United States," *Marquette Law Review* 95 (2011): 53–95.

5 Many have written about the development and evolution of western water law. The historian Donald Worster's account is particularly accessible and recognizes that prior appropriate doctrine arose in different regions of the West almost simultaneously. D. Worster, *Rivers of Empire: Water, Aridity, and the Growth of the American West* (New York: Oxford University Press, 1985), 87–96.

6   The role of public lands is underemphasized as a factor in prior appropriation. D. J. Pisani, "Enterprise and equity: A critique of western water law in the nineteenth century," *Western Historical Quarterly* 18 (1987): 21–22.

7   For dimensions of canals in Utah and Wyoming, see L. J. Arrington and D. May, "A different mode of life: Irrigation and society in nineteenth-century Utah," *Agricultural History* 49 (1975): 8. Walter Webb famously argued that appropriation doctrine was purely a product of geography; it was a law dictated by the law of nature. But Worster argues that it was a product of the time, Bretsen and Hill hold that it was product of need of capital, and Pisani says it was an effect of predominance of public lands. W. P. Webb, *The Great Plains* (Boston: Ginn, 1931), 431–52; Worster, *Rivers of Empire*, 87–92; S. N. Bretsen and P. J. Hill, "Irrigation institutions in the American West," *UCLA Journal of Environmental Law and Policy* 25 (2006): 283–332; Pisani, "Enterprise and equity," 21–22.

8   Lee's Ferry bears the name of what is most likely a falsely accused but executed Mormon settler, and its history is well worth knowing before any excursion into the Grand Canyon. M. Ghiglieri and T. M. Myers, *Over the Edge: Death in the Grand Canyon* (Flagstaff, AZ: Puma Press, 2012), 497–507.

9   A. B. Murphy, "Territory's continuing allure," *Annals of the Association of American Geographers* 103 (2013): 1212–26. The concept of sovereignty applied to the United States—the individual versus the state—has received considerable philosophical attention. J. B. Elshtain, *Sovereignty: God, State, and Self* (New York: Basic Books, 2008), 152–57.

10   For a review of the Harmon Doctrine, as well as its almost complete and immediate abandonment in international water negotiations, see S. C. McCaffrey, "The Harmon Doctrine one hundred years later: Buried, not praised," *Natural Resources Journal* 36 (1985): 549–90.

11   *Kansas v. Colorado*, 185 U.S. 143 (1902). For a brief review of the case: N. Hundley, *Water and the West: The Colorado River Compact and the Politics of Water in the American West*, 2nd ed. (Berkeley: University of California Press, 2009), 74–76.

12   A. T. Wolf, "International water agreements: Implications for the ACT and ACF," in *Interstate Water Allocation in Alabama, Florida, and Georgia*, ed. J. L. Jordan and A. T. Wolf (Gainesville: University Press of Florida, 2006), 139.

13   The background for the Colorado Compact, the events of the commission, and the intricate details are vast. Hundley's *Water and the West* is the superb history.

14   Hundley, *Water and the West*, 188–89. There are interesting similarities between this commission and the Constitutional Convention—such as seclusion to allow compromise. They also both hinged on a single, broad compromise that allowed for subsequent (previously intractable) negotiations.

15   The additional 1.5 million acre-feet (MAF) is for the lower basin and not for delivery at Lee's Ferry. It remains, in fact, an unresolved issue. Arizona has typically argued that it represents the Gila River. Mexico was not part of the initial negotiations; in 1944 the United States signed a treaty with Mexico that allocated 1.5 MAF per year to Mexico and made it the most senior appropriator.

16   California's agreement to live within 4.4 MAF didn't really happen until 2003, under the Quantitative Settlement Agreement. The Supreme Court interpreted the Boulder Canyon Project Act as Congress allocating the water so that California got 4.4 MAF, but California continued to use much more water than that for most of the century.

17   Arrington and May, "A different mode of life," 3–20.

18   Bretsen and Hill, "Irrigation institutions in the American West," 5–6.

19   This was certainly not the opinion of many western leaders at the time, who thought the federal government should have a limited role and should devolve the public lands to the states. The primary arid lands scientist of the nineteenth century, John Wesley Powell, instead pushed for the federal government to retain ownership and take on the role of building the water storage, delivery, and irrigation works so that small farmers would have a chance of making it in the West. Although states' rights advocates drove Powell out of the political arena, most of his opinions and recommendations were realized in the opening decades of the twentieth century. Ironically, John Elwood Mead, who later became commissioner of the Bureau of Reclamation, had once advocated for local control and efforts in western water development. W. L. Graf, *Wilderness Preservation and the Sagebrush Rebellions* (Savage, MD: Rowman and Littlefield, 1990), 19–30; J. R. Kluger, *Turning Water with a Shovel: The Career of Elwood Mead* (Albuquerque: University of New Mexico Press, 1992), 2, 18–23, 115–29.

20   Arrington and May, "A different mode of life," 10.

Chapter 6: A New Water Market

1   J. F. Kenny et al., *Estimated Use of Water in the United States in 2005* (Washington, DC: U.S. Geological Survey Circular 1344, 2005).

2   At 317 *Strickler v. City of Colorado Springs*, 26 P. 313 (Colo., 1891). This is one of the first cases in Colorado dealing with the legality of purchasing an agricultural water right for municipal use.

3   Summary statistics on national water market transactions are difficult because the vast majority of transactions are informal, bilateral trades that are not managed in a central database. The most comprehensive information available is a

proprietary database kept by WestWater Research: M. Payne and T. Ketellapper, *The 2016 Water Market Outlook: Performance, Growth and Investment Trends in the Water Rights and Water Resource Development Sector* (Phoenix, AZ: WestWater Research, 2016). For Freeport-McMoRan purchase of water, see J. W. Miller, "Copper Miners Pressured by Cost of Water as Mineral Prices Slide," *Wall Street Journal*, May 22, 2014.

4   E. K. Wise, "Tree ring record of streamflow and drought in the upper Snake River," *Water Resources Research* 46, no. 11 (2010).

5   C. W. Stockton and G. C. Jacoby, "Long-term surface water supply and streamflow levels in the Upper Colorado River Basin," *Lake Powell Research Project Bulletin* (1976): 18; C. A. Woodhouse, S. T. Gray, and D. M. Meko, "Updated streamflow reconstructions for the Upper Colorado River Basin," *Water Resources Research* 42 (2006).

6   J. Garner, "Drought in Colorado Is Foreign to New Residents," *Rocky Mountain News*, May 3, 2002.

7   P. C. D. Mill et al., "Stationarity is dead. Whither water management?" *Science* 319 (2008): 573–74.

8   N. Hundley, *Water and the West: The Colorado River Compact and the Politics of Water in the American West*, 2nd ed. (Berkeley: University of California Press, 2009), 302–3.

Chapter 7: Running Water

1   M. Lind, *The Land of Promise: An Economic History of the United States* (New York: HarperCollins, 2012), 46.

2   L. P. Cain, "Raising and watering a city: Ellis Sylvester Chesbrough and Chicago's first sanitation system," *Technology and Culture* 13 (1972): 353–72.

3   J. J. Wallis, "American government finance in the long run: 1790 to 1990," *Journal of Economic Perspectives* 14 (2000): 66–67. For Erie Canal bonds as secure currency, see P. L. Bernstein, *Wedding of the Waters: The Erie Canal and the Making of a Great Nation* (New York: W. W. Norton, 2005), 232–34, 352–54.

4   "Valuation, taxation, and public indebtedness, VII," *Tenth Census of the United States, 1880* (Washington, DC: U.S. Government Printing Office, 1884), 523–26.

5   C. Webber and A. Wildavsky, *A History of Taxation and Expenditure in the Western World* (New York: Simon & Shuster, 1986), 382; R. G. McGrane, *Foreign Bondholders and American State Debts* (New York: MacMillan and Company, 1935), 265–67. Poem is from Webber and Wildavsky, *History of Taxation and Expenditure*, 383.

6   C. Gibson, *Population of the 100 Largest Cities and Other Urban Populations in the United States 1790–1990*; Population Division Working Paper No. 97 (Washington, DC: U.S. Bureau of the Census, 1998).

7   L. Cain, *Sanitation Strategy on a Lakefront Metropolis* (De Kalb: Northern Illinois University Press, 1978), 1–5.

8   J. A. Tarr, J. McCurley, F. C. McMichael, and T. Yosie, "Water and wastes: A retrospective assessment of wastewater technology in the United States, 1800–1932," *Technology and Culture* 25 (1984): 228–30.

9   J. T. Cumbler, *Reasonable Use: The People, the Environment, and the State, New England 1790–1930* (New York: Oxford University Press, 2001), 50–62.

10   Cain, *Sanitation Strategy on a Lakefront Metropolis*, 20–26.

11   Tarr et al., "Water and wastes," 237; in particular see table 1 (p. 238), which synthesizes statistics from the U.S. Bureau of the Census, specifying the mileage of storm sewers, sanitary sewers, and combined sewers.

12   Cain, *Sanitation Strategy on a Lakefront Metropolis*, 26–32.

13   E. O. Jordan, "Typhoid fever and water supply in Chicago," *Journal of the American Medical Association* 39 (1902): 1561–66.

14   J. A. Egan, *Pollution of the Illinois River as Affected by the Drainage of Chicago and Other Cities* (Springfield, IL: Phillips Brothers, 1901), xxvii, xx.

15   S. K. Schultz and C. McShane, "To engineer the metropolis: Sewers, sanitation, and city planning in late-nineteenth-century America," *Journal of American History* 55 (1978): 393, 410.

16   For the evolving dispute between physicians and engineers at the turn of the twentieth century, see generally Tarr et al., "Water and wastes," 243–45. The quote about "more equitable" is from "Sewage Pollution of Water Supplies," *Engineering News* 48 (August 1, 1903): 117. The quote "a true and greatest conservationist" is from "A plea for common sense in the state control of sewage disposal," *Engineering News* 67 (February 29, 1912): 412–13. The quote "sentimentalist" is from "Relations between sewage disposal and water supply are changing," *Engineering News Record* 28 (April 5, 1917): 11–12. For more about 90 percent of wastewater going untreated, see W. L. Andreen, "The evolution of water pollution control in the United States—State, Local, and Federal Efforts, 1789–1972: Part I," *Stanford Environmental Law Journal* 22 (2003): 167.

17   Cain, *Sanitation Strategy on a Lakefront Metropolis*, 64.

18   In fact, the flow of the Chicago River had been reversed as early as 1871. But the reversal of the 1880s was the most significant as a permanent reversal of flow.

19   Cain, *Sanitation Strategy on a Lakefront Metropolis*, 73.

20   Egan, *Pollution of the Illinois River*, xxviii.

21   Cain, *Sanitation Strategy on a Lakefront Metropolis*, 146–48, particularly financial data in appendix 2.

22   D. Cutler and G. Miller, "Water, water everywhere: Municipal finance and water supply in American cities," in *Corruption and Reform: Lessons from America's Economic History*, ed. E. L. Glaeser and C. Goldin (Chicago: University of Chicago Press, 2006), 171–73. In the years between the panic of 1873—when states greatly constrained municipal borrowing—and 1890, many towns financed their waterworks through private franchises. Private capital was used to construct the waterworks, but with extremely attractive rights such as exemption from taxes, eminent domain, and freedom from price regulation; such rights were eventually constrained. After 1875, most franchises gave cities the option to purchase the privately constructed infrastructure at any time, which cities eventually did. L. Anderson, "Hard choices: supplying water to New England towns," *Journal of Interdisciplinary History* 15 (1984): 218–21.

23   Cutler and Miller, "Water, water everywhere," 173–76.

24   J. C. Teaford, *City and Suburb: The Political Fragmentation of Metropolitan America, 1850–1970* (Baltimore, MD: Johns Hopkins University Press, 1979), 79.

25   Wallis, "American government finance in the long run," 70.

Chapter 8: Burning Rivers

1   *Anaerobic* means there is no free oxygen; oxygen is available, but it is not free atmospheric oxygen and is instead CO_2.

2   C. Webber and A. Wildavsky, *A History of Taxation and Expenditure in the Western World* (New York: Simon & Shuster, 1986), 422–24.

3   J. J. Wallis, "American government finance in the long run: 1790 to 1990," *Journal of Economic Perspectives* 14 (2000): 61–82.

4   M. V. Melosi, *The Sanitary City: Environmental Services in Urban America from Colonial Times to the Present*, abridged ed. (Pittsburgh, PA: University of Pittsburgh Press, 2008), 236.

5   Melosi, *The Sanitary City*, 198–202.

6   E. H. Monkkonen, *The Local State: Public Money and American Cities* (Stanford, CA: Stanford University Press, 1995), 116; C. R. Hulten and G. E. Peterson, "Capital stocks: Needs, trends, and performance," *American Economic Review* 74 (1984): 169.

7   The 1972 act was named the Federal Water Pollution Control Act. After being revised in 1977, it was renamed the Clean Water Act; the Clean Water Act is generally taken to refer to the 1972 act.

8   C. E. Colten and P. N. Skinner, *The Road to Love Canal: Managing Industrial Waste before EPA* (Austin: University of Texas Press, 2010), 59–60.

9   G. D. Cooke, ed., *The Cuyahoga River Watershed: Proceedings of a Symposium held at Kent State University* (Kent, OH: Institute of Limnology & Department of Biological Sciences, Kent State University, 1968), 90–91.

10 G. Powell, "Walter B. Jones Memorial Award for Coastal Steward of the Year," NOAA Press Release 2005-R416 (Washington, DC: National Oceanic and Atmospheric Administration), 2005.

11 I am indebted to Terrianne Schulte for guiding me through her earlier work on the League of Women Voters and their role in regulating water pollution, and particularly the role of Edith Chase. T. K. Schulte, "Grassroots at the Water's Edge: The League of Women Voters and the Struggle to Save Lake Erie, 1956–1970" (PhD dissertation, Department of History, State University of New York at Buffalo, 2006). For information on the history of women in urban sanitation, see S. M. Hoy, "Municipal housekeeping: The role of women in improving urban sanitation practices, 1880–1917," in *Pollution and Reform in American Cities, 1870–1930*, ed. M. V. Melosi (Austin: University of Texas Press, 1980), 173–98; J. Price, *Flight Maps: Adventures with Nature in Modern America* (New York: Basic Books, 1999), 62.

12 League of Women Voters, *Lake Erie, Requiem or Reprieve? A Study of Lake Erie Problems* (Cleveland, OH: League of Women Voters, Lake Erie Basin Committee, 1966).

13 Pages 32–42 in *Lake Erie, Requiem or Reprieve?* is as fine an analysis of water quality policy and options as would be found in most reports from contemporary think tanks.

14 A thorough review of the rhetoric and reality around the fire can be found in J. H. Adler, "Fables of the Cuyahoga: Reconstructing a history of environmental protection," *Fordham Environmental Law Journal* 14 (2003): 89–146.

15 C. D. Jacobson and J. A. Tarr, "Ownership and financing of infrastructure: Historical perspectives," Policy Research Working Paper 1466 (Washington, DC: World Bank, 1994), 15.

16 Melosi, *The Sanitary City*, 358.

17 For a review of different nutrient limitations in different societies, see D. Cordell, J. Dranger, and S. White, "The story of phosphorus: Global food security and food for thought," *Global Environmental Change* 19 (2009): 292–305.

18 Vaclav Smil has provided a very readable history of how nitrogen was synthesized as part of war efforts, how it transformed world food production, and how it has affected the environment: V. Smil, *Enriching the Earth: Fritz Haber, Carl*

Bosch, and the Transformation of World Food Production (Cambridge, MA: MIT Press), 2001.

19   R. E. Turner, N. N. Rabalais, and D. Justice, "Gulf of Mexico hypoxia: Alternate states and a legacy," *Environmental Science and Technology* 42 (2008): 2323–27.

20   U.S. Environmental Protection Agency, *The National Water Inventory: A Report to Congress for the 2002 Reporting Cycle* (EPA 841-F-07-003), October 2007.

21   W. B. Hildreth and C. K. Zorn, "The Evolution of the State and Local Government Municipal Debt Market over the Past Quarter Century," *Public Budgeting & Finance*, 2005, 127–53.

22   Moody's Investor Service, "U.S. Municipal Bond Defaults and Recoveries, 1970–2009," *Credit Policy*, February 2010, exhibit 10.

23   A law review of municipal sector interest swaps is found in J. Redmond, "State and local governmental entities: In search of . . . statutory authority to enter into interest rate swap agreements," *Fordham Law Review* 63 (1995): 2177.

24   L. J. Stewart and C. A. Cox, "Debt-related derivative usage by U.S. state and municipal governments and evolving financial reporting standards," *Journal of Public Budgeting, Accounting, and Financial Management* 20 (2008): 466–83; R. Weber, "Selling city futures: The financialization of urban redevelopment policy," *Economic Geography* 86, no. 3 (2010), 251–74.

25   There are many analyses of the Jefferson County finances and events; two in particular provide a chronology along with an extremely useful description of the financial instruments that were used: D. V. Denison and J. B. Gibson, "A tale of market risk, false hope, and corruption: The impact of adjustable rate debt on the Jefferson County, Alabama, Sewer Authority," *Journal of Public Budgeting, Accounting, & Financial Management* 25 (2013): 311–45; M. E. Howell-Moroney and J. L. Hall, "Waste in the sewer: The collapse of accountability and transparency in public finance in Jefferson County, Alabama," *Public Administration Review* 71, no. 2 (2011): 232–42.

Chapter 9: Regulating Power

1   U.S. Department of the Interior, *Proceedings of the Endangered Species Committee* (Washington, DC: U.S. Department of the Interior, January 23, 1979).

2   W. J. Novak, *The People's Welfare: Law and Regulation in Nineteenth-Century America* (Chapel Hill: University of North Carolina Press, 1996), 2.

3   N. Rosenberg, "America's rise to leadership," in *America's Wooden Age: Aspects of Its Early Technology*, ed. B. Hindle (New York: Sleepy Hollow Restorations, 1975), 42–43, 56.

4   D. E. Nye, *Consuming Power: A Social History of American Energies* (Cambridge, MA: MIT Press, 1998), 21–22.

5   B. Hunter, "Wheat, war, and the American economy during the age of revolution," *William and Mary Quarterly* 62 (2005): 506, 508–9, 514–16.

6   This suite of characteristics, roles, and responsibilities, evolving through time, was encapsulated by the term *regulation*. The entity is subjected to a constraint on its potential use of its facilities, and thus regulated, results in the key characteristics of what are considered a public utility. L. S. Hyman et al., *The Water Business: Understanding the Water Supply and Wastewater Industry* (Vienna, VA: Public Utilities Reports, 1998), 171–73.

7   D. M. Gold, "Eminent domain and economic development: The Mill Acts and the origins of laissez-faire constitutionalism," *Journal of Libertarian Studies* 21 (2007): 101–22, 104.

8   P. M. Malone, *Waterpower in Lowell: Engineering and Industry in Nineteenth-Century America* (Baltimore, MD: Johns Hopkins University Press, 2009), 25–42.

9   Malone, *Waterpower in Lowell*, 39–41; T. Steinberg, *Nature Incorporated: Industrialization and the Waters of New England* (Cambridge: Cambridge University Press, 1991), 86–88.

10 Malone, *Waterpower in Lowell*, 48.

11 M. J. Horowitz, "The transformation in the conception of property in American law, 1780–1860," *University of Chicago Law Review* 40 (1978): 254–55.

12 Hunter, "Wheat, war, and the American economy," 506–7.

13 Nye, *Consuming Power*, 45–47.

14 Papermills and sawmills were widely seen as not having public purpose, and many legislatures and courts would not allow the owners of such mills to take advantage of mill act regulation or to throw a gristmill into a paper or saw mill operation; Gold, "Eminent domain and economic development," 112.

15 Maine case: *Jordan v. Woodward* 40 Me. 317, 1855. Vermont case: *Williams v. School District No. 6*, 33 Vt. 271, 1860. Broader rejection of mill acts are described in Gold, "Eminent domain and economic development," 117.

Chapter 10: The Power of a River

1   Many inventories of power capacity show steam increasing dramatically in the second half of the nineteenth century; but much of this steam power capacity was actually reserve capacity rather than power used. W. D. Devine, "From shafts to wires: Historical perspective on electrification," *Journal of Economic History* 43 (1983): 351, 369–70.

2   L. Philipson and H. L. Willis, *Understanding Electric Utilities and De-Regulation*, 2nd ed. (Boca Raton, LA: CRC Press, 2006), 82–83.

3   Philipson and Willis, *Understanding Electric Utilities*, 1–13.

4   L. S. Hyman et al., *The Water Business: Understanding the Water Supply and Wastewater Industry* (Vienna, VA: Public Utilities Reports, 1998), 133–35.

5   G. Tollefson, *BPA and the Struggle for Power at Cost* (Portland, OR: Bonneville Power Administration, 1987), 78–83.

6   W. Wells, "Public power in the Eisenhower administration," *Journal of Policy History*, 20, 2008, 227–62.

7   W. J. Hausman and J. L. Neufeld, "Falling water: The origins of direct federal participation in the U.S. electric utility industry, 1902–1933," *Annals of Public and Cooperative Economics* 71 (1999): 49–74.

8   S. M. Neuse, *David Lilienthal: The Journey of an American Liberal* (Knoxville: University of Tennessee Press, 1996), 112.

9   D. E. Lilienthal, "The power of governmental agencies to compel testimony," *Harvard Law Review* 36 (1926): 694–724; D. E. Lilienthal, "Needed: A new railroad labor law," *The New Republic*, 1924: 169–71; D. Lilienthal, "The regulation of public utility holding companies," *Columbia Law Review*, 1929, 408, 404–40; D. Lilienthal, "Recent developments in the law of public utility holding companies," *Columbia Law Review*, February, 1931, 189–207.

10   Description of Lilienthal's time in Wisconsin: Neuse, *David Lilienthal*, 47–58. "Regulation with a vengeance" quote is from Neuse, *David Lilienthal*, 54.

11   Franklin D. Roosevelt, "Message to Congress Suggesting the Tennessee Valley Authority," April 10, 1933, *FDR Presidential Library and Museum*.

12   Neuse, *David Lilienthal*, 77–82.

13   *Tennessee Electric Power Co. v. Tennessee Valley Authority*, 206 U.S. 118 (1939).

14   Willkie's frustration was based on the federal subsidies built into the TVA, which undermined the comparability of its power costs with those of the private sector. The financial basis for the TVA power has been a source of substantial economic analysis, and some analyses are particularly quantitative, systematic, and damning: W. U. Chandler, *The Myth of the TVA: Conservation and Development in the Tennessee Valley, 1933–1983* (Cambridge, MA: Ballinger Publishing Company, 1984), 87–96. Other chapters in Chandler likewise deconstruct many fiscal assertions of the TVA's other functions. Willkie quotes are from Neuse, *David Lilienthal*, 112. Negotiations are described in Neuse, *David Lilienthal*, 109–13.

15   Use of the term *yardstick* by Lilienthal and FDR evolved over time in different contexts: Neuse, *David Lilienthal*, 84–86.

16   The word *authority* (e.g., in TVA) denotes a generation and transmission utility

owned by the government; the TVA is the largest such utility. A *power adminis-tration* (e.g., the Bonneville Power Administration), is a government agency that does not own generation facilities but sells or manages power when produced by other governmental resources, such as the Bureau of Reclamation or the Corps of Engineers. For instance, in the TVA region the Corps of Engineers operates nine hydropower plants, through which it produces 900 MW of power. That power is marketed through the Southeastern Power Administration. Likewise, the Western Area Power Administration markets the power produced by Recla-mation's Glen Canyon Dam.

17   Energy Information Administration, "History of the U.S. electric power indus-try, 1882–1991," in *The Changing Structure of the Electric Power Industry, 2000* (Washington DC: EIA, 2000), 114.

18   W. B. Wheeler and M. J. McDonald, *TVA and the Tellico Dam: A Bureaucratic Crisis in Post-Industrial America* (Knoxville: University of Tennessee Press, 1986), 156–57.

19   The TVA had from its inception been conceived as a unified plan to optimize the benefits for water, power, environment, and navigation; the Tellico Dam was a part of this plan for "unified development" of the watershed that was the signature of the TVA to water managers worldwide. Board of Directors of the Tennessee Valley Authority, *Report to the Congress on the Unified Development of the Tennessee River System* (Knoxville: Tennessee Valley Authority, March 1936).

20   The TVA and the Tellico Dam have received considerable historical and legal analysis. For analysis of how the situation fit into the bureaucracy of the TVA as it was trying to redefine its mission, see Wheeler and McDonald, *TVA and the Tellico Dam.* For a more direct, first-person analysis from the attorney who led the case against TVA, with particular emphasis on the impact to farmers of the valley, see Z. J. B. Plater, *The Snail Darter and the Dam: How Pork-Barrel Politics Endangered a Little Fish and Killed a River* (New Haven, CT: Yale University Press, 2013).

21   U.S. Department of the Interior, *Proceedings of the Endangered Species Committee* (Washington, DC: U.S. Department of the Interior, January 23, 1979), 26.

22   Congressional Research Service, *A Legislative History of the Endangered Species Act of 1973, as Amended in 1976, 1977, 1978, 1979, and 1980* (Washington, DC: U.S. Government Printing Office, 1982), 1292.

Chapter 11: Channelization

1   S. A. Schumm and H. R. Khan, "Experimental study of channel patterns," *Geo-logical Society of America Bulletin* 83 (1972): 1755–70.

2   This is a well-studied phenomenon; for Thompson's own work on pools and rif-
     fles, see D. M. Thompson and K. S. Hoffman, "Equilibrium pool dimensions
     and sediment-sorting patterns in coarse-grained New England channels," *Geo-
     morphology* 38 (2001): 301–16.

3   J. S. Van Cleef, "How to restore our trout streams," *Transactions of the American
     Fisheries Society* 14 (1885): 50–55.

4   E. Van Put, *The Beaverkill: The History of a River and Its People* (New York: Lyons
     & Burford, 1996), 30–35.

5   E. H. Shor, R. H. Rosenblatt, and J. D. Isaacs, *Carl Leavitt Hubbs: 1894–1979:
     A Biographical Memoir*, Vol. 56 (Washington, DC: National Academies Press,
     1987), 241–19.

6   C. Hubbs, C. M. Tarzwell, and J. R. Greely, *Methods for the Improvement of Mich-
     igan Trout Streams* (Ann Arbor: University of Michigan Press, 1932); H. Clep-
     per, *Origins of American Conservation* (New York: Ronald Press Company, 1966),
     64–68.

7   These scientists employed techniques that remain the standard for evaluating
     the efficacy of restoration actions—the Before, After, Control, Intervention
     study design. D. S. Shetter, O. H. Clark, and A. S. Hazzard, "The effects of
     deflectors in a section of a Michigan trout stream," *Transactions of the Amer-
     ican Fisheries Society* 76 (1949): 248–78. A good example of this intervention
     study design applied to restoration is the work and career of Clarence Tarzwell,
     who went from the University of Michigan program to the U.S. Forest Ser-
     vice in Albuquerque and then to the Tennessee Valley Authority, yet applied
     the training he received at the Michigan School throughout: C. M. Tarzwell,
     "Experimental evidence on the value of trout stream improvement in Michigan,"
     Transactions of the American Fisheries Society 66 (1937): 177–87. Quote is from
     T. K. Chamberlain and W. W. Huber, "Ten years of trout stream management on
     the Pisgah," *Progressive Fish Culturalist* 9 (1947): 185–91.

8   D. M. Thompson, "Did the pre-1980 use of in-stream structures improve streams?
     A reanalysis of historical data," *Ecological Applications* 16 (2006): 784–96.

9   D. M. Thompson, *The Quest for the Golden Trout: Environmental Loss and Ameri-
     ca's Iconic Fish* (Hanover, NH: University Press of New England, 2013), 186–97.

10   For New York, see U.S. Department of Agriculture, *State Forests for Public Use*
     (Washington, DC: U.S. Government Printing Office, 1940). For California, see
     R. Ehlers, "An evaluation of stream improvement devices constructed eighteen
     years ago," *California Fish and Game* 42 (1956): 203–17. For Wyoming, see J. W.
     Mueller, "Wyoming stream improvement," *Wyoming Wild Life* 18 (1954): 30–32.
     Thompson provides an excellent review of the early years of in-stream resto-

ration people and projects. D. M. Thompson, "The history of the use and effectiveness of instream structures in the United States," in *Humans as Geologic Agents*, ed. J. Ehlen, W. C. Haneberg, and R. A. Larson (Boulder, CO: Geological Society of America Reviews in Engineering Geology, 2005).

11   Quote is from Chamberlain and Huber, "Ten years of trout stream management on the Pisgah," 185. In fact, the public fishing easements purchased along selected trout streams of the Catskills were intended primarily to provide the Conservation Department with another opportunity to do stream improvement work. Van Put, *The Beaverkill*, 256-57.

12   J. H. Thorp, M. C. Thoms, and M. D. Delong, "The riverine ecosystem synthesis: Biocomplexity in river networks across space and time," *River Research and Applications* 22 (2006): 123-47.

13   A. Brookes, *Channelized Rivers: Perspectives for Environmental Management* (Chichester, UK: John Wiley & Sons, 1988), 10, 18-19. National Research Council, *Restoration of Aquatic Ecosystems* (Washington, DC: National Academies Press, 1992), 194.

14   J. L. Funk and C. E. Ruhr, "Stream channelization in the Midwest," in *Stream Channelization: A Symposium*, ed. E. Schneberger and J. L. Funk (Omaha, NE: North Central Division of the American Fisheries Society, Special Publication No. 2, 1971), 10.

15   C. J. Barstow, "Impact of channelization on wetland habitat in the Obion–Forked Deer Basin, Tennessee," in Schneberger and Funk (ed.), *Stream Channelization: A Symposium*, 23; D. R. Hansen, "Stream channelization effects on fishes and bottom fauna in the Little Sioux River, Iowa," in Schneberger and Funk (ed.), *Stream Channelization: A Symposium*, 41.

16   Committee on Government Operations Report, *Stream Channelization: What Federally Financed Draglines and Bulldozers do to Our Nation's Streams*, Fifth Report by the Committee on Government Operations, House Report No. 93-530 (Washington, DC: U.S. Government Printing Office, 1973), 7.

17   Brookes, *Channelized Rivers*, 20.

18   Federal Highways Administration, *Restoration of Fish Habitat in Relocated Streams: Federal Highways Administration Report No. FHWA-IP-79-3* (Washington, DC: U.S. Department of Transportation, 1979).

19   C. H. Pennington et al., *Biological and Physical Effects of Missouri River Spur Dike Notching: Army Engineer Waterways Experiment Station Report ADA199779* (Vicksburg, MS: U.S. Army Corps of Engineers, 1988).

20   M. J. Migel, *The Stream Conservation Handbook* (New York: Crown Publishers, 1974), 121-22.

21   M. Ondaatje, *The English Patient* (London: Bloomsbury, 1992), 261–62.

22   L. B. Leopold, J. P. Miller, and G. M. Wolman, *Fluvial Processes in Geomorphology* (San Francisco: Freeman, 1964).

Chapter 12: The Restoration Economy

1   E. S. Bernhardt, et al., "A nationwide synthesis of stream restoration," *Science* 308 (2005): 636–37.

2   J. McPhee, *Encounters with the Archdruid* (New York: Ballantine Books, 1971), 158–59.

3   E. H. Stanley and M. W. Doyle, "Trading off: The ecological effects of dam removal," *Frontiers in Ecology and the Environment* 1 (2003): 15–22.

4   D. Yergin and J. Stanislaw, *The Commanding Heights: The Battle for the World Economy* (New York: Free Press, 1998), 123–31.

5   J. H. Dales, *Pollution, Property and Prices: An Essay in Policy-making and Economics* (Toronto: University of Toronto Press, 1968).

6   T. E. Dahl, *Wetland Losses in the United States, 1780s to 1980s* (Washington, DC: U.S. Fish and Wildlife Service, 1990); P. Hough and M. Robertson, "Mitigation under Section 404 of the Clean Water Act: Where it comes from, what it means," *Wetlands Ecology and Management* 17 (2009): 15–33.

7   M. W. Doyle and F. D. Shields, "Compensatory mitigation for streams under the Clean Water Act: Reassessing science and redirecting policy," *Journal of the American Water Resources Association* 48 (2012): 494–509.

8   For road-building, land development, and stream and wetland credit summaries, see T. K. BenDor, J. A. Riggsbee, and M. W. Doyle, "Risk and markets for ecosystem services," *Environmental Science and Technology* 45 (2012): 10322–30.

9   J. Ott, "'Ruining' the rivers of the Snake Country: The Hudson's Bay Company's Fur Desert Policy," *Oregon Historical Quarterly* 104 (2003): 166–95; D. Muller-Schwarze, *The Beaver: Its Life and Impact*, 2nd ed. (Ithaca, NY: Cornell University Press, 2011), 160–64.

INDEX

Page numbers in *italics* refer to illustrations.